REDISCOVERED CLASSICS

THE ART OF TYING KNOTS

EDITED BY PAUL N. HASLUCK

WITH A FOREWORD BY DES PAWSON

ENDLESS MOUNTAINS
PUBLISHING COMPANY
MOOSIC, PENNSYLVANIA

ISBN: 978-0-9988527-3-7

This is a reproduction of the 1908 edition of
Knotting and Splicing Ropes and Cordage
edited by Paul N. Hasluck

With a new foreword by Des Pawson

Published by Endless Mountains Publishing Company
Moosic, Pennsylvania
www.EndlessMountainsPublishing.com

MANUFACTURED IN THE UNITED STATES OF AMERICA

FOREWORD

Knots are the building blocks of civilization. Even so, few books were written about them until the end of the 19th Century. Before that, knots and splices were just touched upon in seamanship works, some military works., and a few magazine articles. The knowledge of most knots and splices was passed from hand to hand, shared down the generations.

When this book was first published in the United Kingdom in 1904, it was the most comprehensive work on knots available. It was highly successful, remaining in print almost continuously since its earlier days, and here presented in a facsimile of the American 1908 edition, which was the same as the British edition.

Paul Nooncre Hasluck (1854–1931) was an Australian editor and author of many popular practical books aimed, like this one, at a general audience. He was a pioneer in the "do-it-yourself" publishing category. This book originated from a series of articles written by Lancelot L. Haslope and edited by P. N. Hasluck, that were published between 1891 and 1892 in *Work,* an illustrated Magazine. Though the author pulled from a wide variety of sources for this book, it is still very much of its day in that the majority of sources were of a maritime nature. To complete the book Hasluck has added some extra material.

Perhaps to broaden the appeal beyond seafarers, non-maritime names have been used. For example, the illustration of what is commonly defined as the Clove Hitch is here called a Builder's Knot, just putting the name Clove Hitch in the body of the text. He does much the same with that most ancient of knots, the Sheet Bend, describing the weaver's method of tying and giving it the name Weaver's Knot. He does go on to show why this structure is called the Sheet Bend and illustrates the Double Sheet Bend, a very useful method of joining together cordage of all types, including the often more slippery synthetic materials common today.

The book has some curious inclusions. For example, there are plenty of Sheepshanks, a fairly useless arrangement. This can be explained by the fact that the Sheepshank was included in the very first seamanship works, perhaps because of its curious and memorable name, and is to be found in every work of knots subsequently, indeed right up until today.

There are a few points where instructions need updating for today's more slippery synthetic ropes. For example, in the section on splicing, it's best to make extra tucks. The advice on splicing the cotton ropes used in the Lancashire mills, shows a real understanding of the problem of splicing cotton rope, which tends to lose its lay. His suggestion of practicing with the much better-behaved hemp rope, could be followed still today, as many of the synthetics behave much the same as the cotton rope.

When writing about fancy knots, good advice is given in the introduction, to make up a bundle of small line to practice Wall Knots, Crown Knots and their various combinations. That advice is as valid today as it was at in Hasluck's day.

Unfortunately, like so many others before, his instructions

for making a Turk's Head originate from D'Arcy Lever's *Young Sea Officer's Sheet Anchor,* published in 1808 and, whilst it is just possible to make a Turk's Head if you know what you are aiming at, I am really not sure that anyone ever actually tried to follow the instructions.

Hasluck's comprehensive section on working with wire rope is a very useful insight into the way in which this most difficult of cordages was handled at the time of writing and has a few pointers still of interest today.

This work contains many of the knots that are needed in a basic knot repertoire: the Figure-of-eight Knot, the Clove Hitch, the Bowline, the Fisherman's Knot, the Fisherman's Bend, Wall and Crown Knots and perhaps makes us look again at some knots that have been overlooked over the years.

Knots are notoriously difficult to draw and the illustrator can frequently get something wrong, as in the example of the Figure-of-eight Knot, which was incorrectly shown in the very first edition, but has been corrected by the time of this 1908 edition. Knots are tools and care should be taken when using them, especially if using some of today's slick synthetic materials.

This book, built on books from the past, has gone on to be one of the building blocks of knot knowledge for later generations of knot books. It contains useful information and insights to a time past. Whether your interest is nautical, historical, mechanical, or crafts, there's much to discover—or rediscover—in this classic collection on the art of tying knots.

—Des Pawson MBE, *March 2018*

CONTENTS.

PREFACE.

THIS Handbook contains, in a form convenient for everyday use, a comprehensive digest of the information on Knotting and Splicing Ropes and Cordage, scattered over more than twenty thousand columns of WORK—one of the journals it is my fortune to edit—and supplies concise information on the details of the subjects on which it treats.

Readers who may desire additional information respecting special details of the matters dealt with in this Handbook, or instructions on kindred subjects, should address a question to The Editor of WORK, La Belle Sauvage, London, E.C., so that it may be answered in the columns of that journal.

P. N. HASLUCK.

La Belle Sauvage, London.

LIST OF ILLUSTRATIONS.

KNOTTING AND SPLICING ROPES AND CORDAGE.

CHAPTER I.

INTRODUCTION: ROPE FORMATION.

KNOTTING is an ancient device with which very early inhabitants of this earth must have been acquainted. From the beginning, mankind must always have used some kind of knot to join animal sinews, plant fibres, or hide strips which, in ancient days, were the prototypes of the varieties of cordage now employed.

A large number of knots has been invented by the skill of man, and on their strength and correct tying depend the lives of thousands and thousands of workmen—seamen, building trade operatives, etc., day by day. The importance of being able to make the knot best suited for the occasion both rapidly and correctly may come in a new light to some when it is pointed out that both lives and property have over and over again been sacrificed to ill-made knots; and this little volume is put forward in the belief that few things better repay the workman's time and trouble in learning than the manipulation of ropes and cordage.

Cordage is used almost daily by everyone in some form or other, but comparatively few can handle it methodically. Men break their nails and teeth gnawing at their own knots endeavouring to untie them, and time and material are wasted. Time spent in learning a few of the simple bends and hitches, reliable under strain and easy to unbend when the strain is released, would never be regretted. It is not necessary for a landsman to learn all the numerous uses to which rope is put, but a knowledge of common "bends" is an inestimable convenience, if not a necessity.

The security of a knot ought not to be, as many seem to think, in the number of turns or hitches in its composition, but in the efficacy of the nip. A "bend" or "hitch" must be so formed that the part of rope under strain nips some portion of the knot, either against itself or the object to which it is attached; and in learning a bend, or impressing it upon one's memory, it will be found most helpful to notice particularly the nip of each separate one as it is studied.

Rope, though usually of hemp, is made of other materials for certain purposes. Coir rope (cocoanut fibre), being light and buoyant, is useful for warps, rocket lines, life-buoy lines, and drift-nets. Manilla grass is adapted for reef points, yachts' hawsers, and wherever tar would be injurious. Hide is required for wheel-ropes, or where great strength with pliability and small circumference is needed. Cotton is serviceable for fancy work, etc. The "yarns" are formed by twisting the hemp right-handed; the "strands," by twisting or laying up the yarns left-handed; and the rope, by laying up the strands right-handed.

Three ropes laid up left-handed form what is known as a cable-laid rope; four-stranded ropes are laid round a heart. Ropes are sometimes laid left-handed, but if the strands are to be laid left-handed the yarns are laid right-handed. If the parts of hemp, etc., be twisted more than is necessary to hold them together, strength is lost. Upon following the course of a yarn in a rope it will be found that, by this alternate laying, it runs nearly straight with the direction of the rope's length.

A three-stranded rope will bear a greater strain

in comparison with its size than any other of the same material; cable-laid ropes and four-stranded ropes are, roughly speaking, about one-fifth weaker. Rope is measured by its circumference, and is laid up in lengths of 113 fathoms, sizes varying up to 28 in.; but it is not usually made up in coils when the size exceeds 5 in. Very small ropes are distinguished by their yarns rather than their size; thus sailors speak of nine-, twelve-, and eighteen-yarn stuff, which is commonly called "seizing stuff."

If the fibres of which a rope is composed were laid parallel to one another and fastened at the two ends, the combined strength of these fibres would be utilised to the full; in other words, they lose strength by being twisted or "laid up." But, on the other hand, the length of the fibres being at most but a few feet, their usefulness in this state is very limited, and the inconvenience of using them so is prohibitive. For this reason the fibres are first twisted into "yarns"; these, again, are laid up into "strands," a strand being formed of several yarns; and, finally, three or more strands are formed into a rope. As twisting diminishes the strength of a rope, it is important that the yarns be carefully laid up, so as to bring an even strain on every part. It should not be laid up too hard—that is, it should only have sufficient twist in it to prevent the fibres from being drawn out without breaking.

"Hawser-laid" ropes are made of three strands laid right-handed, or "with the sun," as it is termed. "Shroud-laid" are made of four strands laid right-handed. A "cable-laid" rope is made of three hawser-laid ropes laid up left-handed, and therefore contains nine strands. Obviously the size of a rope is regulated by the quantity of yarns composing the strands, and not by the number of strands that it contains.

With regard to the weight of ropes, it may be said that ropes of all kinds are usually measured by their circumference. The weight of clean, dry, hemp rope in pounds per fathom is one-fourth of the square of the circumference in inches; for example, a 3-in. hemp rope (about 1 in. in diameter) weighs $\frac{1}{4} \times 3^2 = 2\frac{1}{4}$ lb. per fathom (6 ft.). A flat hemp rope, with a width of about four times the thickness, weighs in pounds per fathom about twice the square of the circumference in inches; for example, a 3-in. by $\frac{3}{4}$-in. flat hemp rope will weigh about $2 \times 7 = 14$ lb. per fathom.

Round wire ropes weigh in pounds per fathom seven-eighths of the square of the circumference in inches; for example, a 3-in. wire rope weighs about $\frac{7}{8} \times 3^2 = 7\frac{7}{8}$ lb. per fathom. A flat wire rope weighs in pounds per fathom ten times the sectional area in square inches; for example, a flat wire rope, 3 in. by $\frac{3}{4}$ in. = say 2 sq. in. area, will weigh about $10 \times 2 = 20$ lb. per fathom.

The maximum safe load on a rope depends on many circumstances, such as quality, age and dryness of rope, nature of load, mode of lifting, etc. Approximately, the safe load on a new hemp rope in hundredweights with direct lift is three times the weight in pounds per fathom. On a sound old rope fall one-half the square of the circumference is sufficient load. A Bessemer steel wire rope will safely carry in hundredweights three times the square of its circumference in inches, and a crucible steel wire rope four times the square of its circumference. For hemp ropes the minimum diameter of sheave should be circumference of rope + 2, and for wire ropes the diameter of sheave in inches should be equal to circumference of rope in sixteenths.

The principle of rope making is very readily shown by holding the ends of a piece of twine or whipcord, about a foot long, in the hands and twisting it so as to increase the lay. If the twine be now slackened by bringing the hands nearer to one another, a loop will first form in the middle of the twine, and it will continue to twist itself up into a compact cord which will not unlay, as the tension to which the strands have been subjected causes friction between them, which holds them together. In other words, the tendency of each part singly to unlay, acting in opposite directions, is the means of keeping them together when joined.

Some very interesting experiments were made by Réaumur, the purposes of which were to ascertain the loss of strength occasioned by laying up the fibres of various substances, one or two of which are given.

1. A thread, consisting of 832 fibres of silk, each of which carried 1 dram and 18 grains, broke with a weight of 5 lbs., though the sum of the absolute strength of the fibres is 104 drams, or upwards of 8 lbs. 2 oz.

2. Three threads were twisted together, their mean strength being nearly 8 lbs. They broke with 17½ lbs., whereas they should have carried 24 lbs.

These experiments prove that though convenience and portability are gained by twisting the fibres, there is a great loss in the strength of the resultant rope.

In speaking of the size of a rope, the circumference and not the diameter is alluded to. Thus, a three-inch rope would be slightly less than an inch in diameter.

In practising knotting it is as well to use a tolerably firm material, such as whipcord, for small common knots, or, still better, line used for sea fishing. Either can be tied up and undone over and over again without injuring it, which is not the case with twine; it is also more easy to see which way the parts of a knot lie in the harder material, and then to find out whether the turns are properly made or not. For more complicated knots, particularly those where the strands of the rope have to be unlaid to form the knot, such as a wall knot (p. 66) or a Matthew Walker (p. 70), it is advisable to use three strands of fishing line, each about a foot long. If a "seizing" (a seizing is shown in Fig. 57, p. 51) be put round them in the centre, so as to hold them firmly together, a good representation of a rope with the strands unlaid ready for working is obtained. A knot can be made and unmade as often as required in this way, without detriment to the strands; but the strands of a rope, owing to their loose nature, will seldom bear knotting more than once or twice. If desired, the knots can be made as above described and kept for future reference. In string also it is better to use hard laid stuff at first, but when these matters are thoroughly understood knots can be made on any sort of cordage without difficulty.

CHAPTER II.

SIMPLE AND USEFUL KNOTS.

THE simplest knot that is made is the overhand knot (Fig. 1). It is very useful, and forms a part of many other knots. To make it, the standing part of the rope—that is, the main part in opposition to the end—is held in the left hand, and the

Fig. 1.—Overhand Knot.

end of the rope is passed back over it (whence its name) and put through the loop thus formed. It is used at the end of a rope to prevent the strands unlaying, and sometimes in the middle of a rope as a stopper knot. If the end of the rope is passed through the "bight" or loop two, three, or more times before hauling it taut, the double,

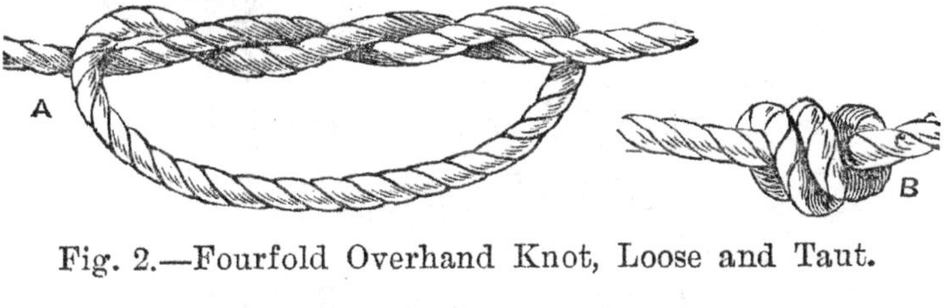

Fig. 2.—Fourfold Overhand Knot, Loose and Taut.

treble, or fourfold knot, A (Fig. 2), is obtained. This is a larger knot than Fig. 1, and is often used on the thongs of whips, being then termed a blood knot. B (Fig. 2) shows the knot hauled taut. Fig.

1 also goes by the name of the Staffordshire knot, as it forms the insignia of the county. A Flemish or figure-of-eight knot is shown by Fig. 3. Tc make it, pass the end of the rope back, over, and

Fig. 3.—Figure-of-eight Knot.

round the standing part, and up through the first bight. The Flemish knot is used for much the same purposes as the preceding knots, but is rather more ornamental.

The bight of a rope is the loop formed when a rope is bent back on itself, in contradistinction to the ends.

The conditions under which the ends of two pieces of cordage have to be joined together are various, and several methods are brought into requisition; but it is always of considerable importance that the most suitable knot be employed in each case. The value of some knots consists

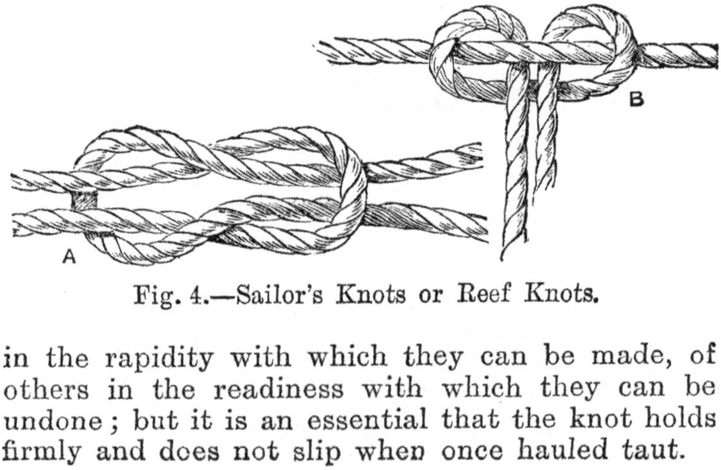

Fig. 4.—Sailor's Knots or Reef Knots.

in the rapidity with which they can be made, of others in the readiness with which they can be undone; but it is an essential that the knot holds firmly and does not slip when once hauled taut.

The commonest knot for joining the ends of two ropes, and probably the knot that is most often made, is the sailor's, true, or reef knot (Figs. 4 and 5). When correctly made it is as perfect as a

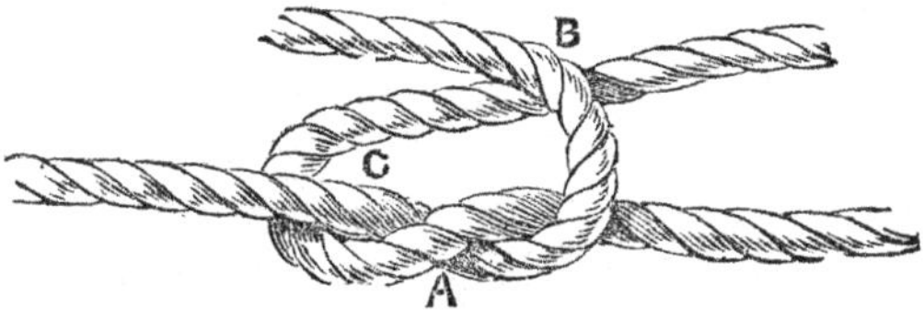

Fig. 5.—Reef Knot, Half-made.

knot can be. It can be made and undone with equal rapidity, and is very secure when taut. Its one disadvantage is that it will not answer when made with ropes of different sizes, as it then slips and comes adrift, but where the two pieces of cordage are of the same size it is most secure and reliable, the strain being equally distributed on every part. It requires a little practice to make it properly. To do this, take an end in each hand and lay one over the other, the right end being undermost; bring the left-hand end under the standing part of the right end, as shown at A (Fig. 5), and over the end at B, round it, and up through the bight at C. The key to the knot is the putting of the right end under the left when the two ends are crossed at the beginning of the

Fig. 6.—Granny or Lubber's Knot.

knot, as the left-hand end then comes naturally first over and then round the other rope, and the ends lie parallel with the standing parts, as at A (Fig. 4).

If the ends are not passed correctly, a granny, lubber's, or calf knot results. This is shown in Fig. 6. Though at first sight this seems to be a good knot, yet it is not so in reality, and when any strain comes upon it it slips and becomes useless. Fig. 7 is a granny knot, as it appears when hauled upon. It is considered a very lubberly thing to make a granny knot, and readers should practise until they can make a true knot rapidly and with certainty in any position.

The sailor's knot is invariably used for reefing sails, the ease with which it can be undone making it very valuable for this purpose. It is only necessary to take hold of the two parts on each

Fig. 7.—Granny Knot, Taut.

side just outside the knot and bring the hands together, and the loops slip over one another, as in Fig. 4, and the knot can be opened at once.

This knot has a curious peculiarity which is not generally known. If the end of one of the ropes is taken in one hand and the standing part of the same rope in the other, and both are hauled until the rope is straight, the knot becomes dislocated, so to speak, and the rope not hauled upon forms a hitch, B (Fig. 4), round the other part. This property was the secret of Hermann's celebrated trick, "the knotted handkerchiefs." After the handkerchiefs, knotted together at the corners, were returned to him by the audience, under pretence of tightening the knots still more, he treated each knot as has been described. The knots seemed firm, but really were loosened so that a touch with his wand separated them easily.

The common bow or rosette knot is a modification of the sailor's knot. The first part of the process of making it is the same, but instead of passing one end singly over and under the other, as in the sailor's knot, both ends are bent back on themselves, and the double parts worked as before. Care must be taken to pass these doubled

Fig. 8.—Overhand Rosette Knot or Bow.

ends exactly as those described in the sailor's knot, or a granny bow will result. Some persons' shoes always come untied, the reason being that they are tied with granny instead of true bows.

Another way of joining the ends of two pieces of cordage is shown in Fig. 8. This is merely an overhand knot, made with two ropes instead of one. Sometimes it is called an openhand knot. It can be made very quickly, and there is no fear of its slipping, but if there is much strain put upon it the rope is very apt to part at the knot, in consequence of the short "nip," or turn, that it makes just as it enters the knot.

Fig. 9 shows the weaver's knot partly made, and Fig. 10 the same knot completed, but not hauled taut. Weavers call this the "thumb knot," as it is made over the thumb of the left hand, and is used by them in joining their "ends" as they break. The rapidity with which they make the knot, snip off the ends, and set the loom going again is wonderful. Netters use this knot to join their twine, and it also forms the mesh of the netting itself, though, of course, it is then made in a very different way. In making the

Fig. 9.—Weaver's Knot, Half-made.

weaver's knot, the two ends to be joined are crossed in the same way as in the sailor's knot, placing the right end under, and holding them with the thumb and finger of the left hand at the place where they cross. The standing part of the right-hand rope is then brought back over the thumb and between the two ends, as shown in Fig. 9. The end A is then bent down over it, and held with the left thumb, while the knot is completed by hauling on B.

An excellent way of joining two ropes is

illustrated by Fig. 11. The ends are laid alongside one another, overlapping each sufficiently to give room for the knot to be made. The double parts are then grasped in each hand and an overhand knot is formed, which is made taut by hauling on both parts at once, as if the knot were single.

Fig. 10.—Weaver's Knot, Closed.

Though the above is the easiest way to make the knot, it is not available where the ropes are fast. In this case a simple knot is made on the end of one rope, but not drawn taut. The end of the other rope is passed through the bight of the first, and a second loop formed with it alongside the first. The knot is closed by drawing the two ropes as before. This is in every way an excellent knot, and very secure.

Fig. 12 shows the ends of two ropes joined by means of a Flemish knot. It does not require

Fig. 11.—Overhand Knot Joining Two Ropes.

much description, and is made after the manner of the knot last described.

The fisherman's knot (Fig. 13) derives its name from the fact that it is always used for joining silkworm gut for fishing purposes. In making it, the strands are made to overlap one another, and an overhand knot is made with one end round the other strand. The strands are turned round, and

Fig. 12.—Flemish Knot Joining Two Ropes.

another overhand knot made with the other end round the first strand. When the knot is tightened

Fig. 13.—Fisherman's Knot.

by hauling on the standing parts, one knot jams against the other and holds securely. The knot is improved by putting the ends twice through their respective loops, as at A (Fig. 2, p. 15). The size of the knot is increased by this means, but it will stand a much heavier strain, so that it is advisable to do this whenever the size of the knot is not of paramount importance.

The whipcord knot (Fig. 14) is used to fasten the lash to a whip. The lash B is laid across the ends of the thong A which are turned up over it.

Fig. 14.—Whipcord Knot.

The lash is brought completely round the thong and through the loop it makes, which secures the ends of the thong firmly. If a silk lash is used, the short end is cut off, but if whipcord, the two ends are generally twisted together for a few inches, as at B, and an overhand knot made with one end round the other, to secure them. The remaining part is left somewhat longer, and another overhand knot at the end prevents it from unravelling

CHAPTER III.

EYE KNOTS, HITCHES, AND BENDS.

ONE of the simplest eye knots is shown by Fig. 15, and is known as the "running" or "slip knot." A bight is first formed, and an overhand knot made with the ends round the standing part. The last named may be drawn through the knot, and the eye made to any size required. There is less chance of the knot coming undone if an overhand

Fig. 15.—Running Knot.

knot is made on the end A. With this knot a sailor ties his neck-handkerchief.

Fig. 16 is the "fisherman's eye knot." A bight is first made of sufficient length, and an overhand knot formed with the standing part round the other strand; the end is now passed round the standing part, and knotted as before. Thus there is a running knot A, with a check knot B, which, when hauled upon, jam tight against one another, and hold securely. This is one of the best knots for making an eye in fishing, as the strain is divided equally between the two knots.

A common way of making an eye on the end of a piece of cord is illustrated by Fig. 17. It is practically the same knot as Fig. 8 (p. 19), except

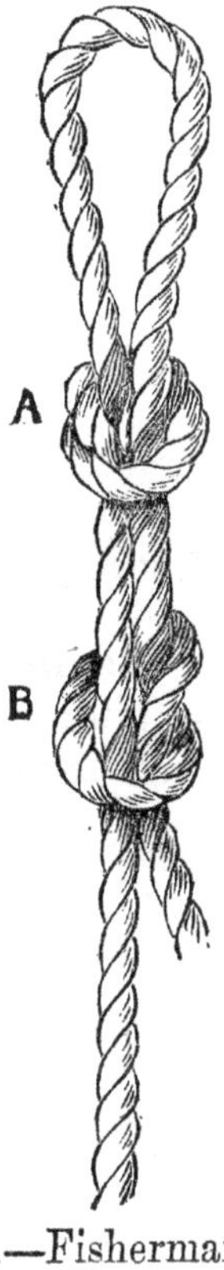

Fig. 16.—Fisherman's Eye Knot.

Fig. 17.—Openhand Eye Knot.

that only one rope is used. The end is brought back along the rope to form the eye, and an overhand knot made with the two parts. This knot, from being so easily made, is often used, but it lacks strength, like the openhand knot (Fig. 8), and should not be used where it is required to bear much strain. It will have been noticed how very often openhand knots form the component parts of other knots.

Fig. 18.—Flemish Eye Knot.

Fig. 18 is an eye made with a "Flemish" knot. It is worked just the same as a single Flemish knot (Fig. 12, p. 22), the only difference being

Fig. 19.—Crabber's Eye Knot.

that two parts are used instead of one. It is stronger, but clumsier, than the one just described, and is not much used.

The "crabber's knot" (Fig. 19) is a curious and not very well known knot, but it is unlikely to part when strained. To make it, bring the end back to form a loop, taking it first under and then over the standing part, up through the main loop, over the standing part again, and up through its own bight. Before the turns are hauled into their places, the knot will slip on the part A, as

Fig. 20.—Bowline Knot.

in an ordinary slip knot; but if the part B is hauled upon, the strand A, which passes through the centre of the knot, rises, and the coil which goes round it jams, making the knot secure: so that it may be used as a running knot or otherwise, as desired. This is also called a running knot with crossed ends.

The "bowline knot" (Fig. 20) cannot slip, and is therefore always used for slinging a man for the purpose of doing some particular piece of work;

the workman sits in the sling. The end is first laid back over the standing part, so as to form a loop; the end is then passed up through the loop, round the back of the standing part, and down through the loop again. Hauling on the end and the standing part makes the knot taut.

A modification of this knot, called a "Bowline on a Bight," is shown by Fig. 21. The loop is made as in the previous knot, only with the two

Fig. 21.—Running Bowline on Bight.

parts of a doubled rope; the bight is then passed up through the loop, opened, and turned backwards over the rest of the knot, when it appears as illustrated. To untie it, draw the bight of the rope up until it is slack enough, and bring the whole of the other parts of the knot up through it, when it will readily come adrift. If the standing parts of the rope are held fast, it puzzles the uninitiated to undo it.

A "Running Bowline" has the knot made on the end after it has been passed round the standing part, thus forming a loop through which the main rope will run. Two ropes may be joined together by making a bowline in the end of one of them, and putting the end of the other through the bight, and forming with it another bowline on its own part. This is often used to join hawsers together.

Fig. 22.

Fig. 23.

Figs. 22 and 23.—Running Knot with Two Ends, Loose and Fastened.

Fig. 22 shows a method of making a rope fast to a post or pillar. The rope is doubled and passed round the post, and the ends put through the loop. For greater security, the ends may be passed round the standing part and through the bight thus formed, as in Fig. 23; or, instead of passing the cords through the bight, a loop may be formed by doubling the ends, and this loop put through the bight, thus forming a slippery hitch. This knot has the advantage of being more readily undone than the other one, as it is necessary merely to pull at the ends, and the rope is released at once. The ends may also be secured by making a Flemish knot on them, instead of an overhand knot.

The remainder of this chapter will discuss a different class of fastenings. It is not easy to state, however, where knots end and bends and hitches begin; indeed, a tie that, in certain circumstances and made a particular way, is called a "knot," differently constructed, and under other

Fig. 24.—Two Half Hitches.

conditions, is called a "bend" or "hitch," though the result is the same in both cases. As an illustration, take two half hitches (Fig. 24), which, if made in another way round a pole, are called a "builder's knot." If readers will analyse the knots illustrated throughout this book, they will find several other similar instances. A (Fig. 24) is a single hitch, being merely a loop formed in a rope. This is readily done by holding the rope in the left hand, and giving it a twist with the right; the loop then forms itself, as it were. When a tightly laid piece of cordage is twisted, these loops are apt to rise and form "kinks," which are very objectionable, as the cord is sure to part at the kink when a strain is put on it.

It is still worse in the case of wire, which breaks readily when kinked. Tight, hard cordage should always be well stretched before it is used, to avoid kinking.

Two half hitches (Fig. 24) are a useful knot for a variety of purposes, as they are quickly made, and will not slip, no matter what strain is put upon them—indeed, the more they are hauled upon the faster they hold. They are the best means of making a rope fast to a hook. First one hitch is slipped on, and then the other on the top of it, and the rope is fast in less than two seconds. This knot is used by surgeons in reducing a dislocation of the thumb joint.

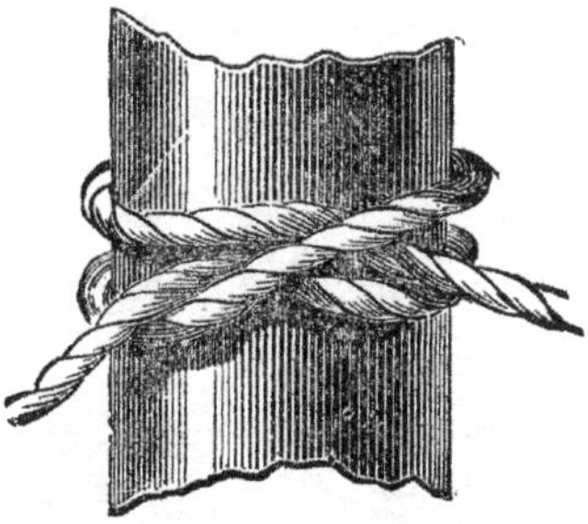

Fig. 25.—Builder's Knot, or Clove Hitch.

Fig. 25 is the builder's knot, merely two half hitches, but as it is used in places where the hitches cannot be passed over the ends of the timber it is made by holding one end in the left hand, passing the rope round the pole, under the end, round the pole again, above the first part, and under its own part; from its non-liability to slip laterally this knot is always used to fasten one pole to another in fitting up scaffolding, from which circumstance it has acquired its name. If, instead of beginning the knot as in Fig. 25, the end is passed, after it has gone round the pole, two or three times round the other part, as in Fig. 2 (p. 15), the remainder of the knot is rather more easily made, as it holds itself taut, and will not slip while the end is put round to complete the fastening.

A "builder's double knot" is made in the same way, except that the end goes round again, as before, and underneath its own part, so making it much stronger. When a builder's knot is made on a rope for the purpose of securing a small line to a stout rope, it is called a "clove hitch."

The "timber hitch" (Fig. 26) is a rough and ready way of securing a piece of timber or any-

Fig. 26.—Timber Hitch.

thing similar; it is made by bringing the end of a rope round the timber, then round the standing part, and then, taking two or more turns, round its own part. The pressure of the coils one over the other holds the timber securely, and the more it is hauled on the tighter it holds. It can be cast off readily.

Fig. 27 is the "killick hitch," a modification of the timber hitch, used for fastening a stone to the end of a rope. After making a timber hitch and hauling it taut, a single hitch is made, and slipped over the end of the stone alongside of it. Some of the best fishing grounds are on rocky coasts where an anchor would not hold; and if it did,

Fig. 27.—Killick Hitch.

there might be considerable risk of losing it altogether, from its jamming in the crevices of a rock. In these places a killick, or large stone, slung as shown in Fig. 27, is used, which holds

Fig. 28.—Magnus Hitch.

the boat by its own weight, without any risk of getting fast to the ground.

The "magnus hitch" (Fig. 28) is a method of securing a rope to a spar, as there is but little tendency to slip endways along the spar. In making it, take the and of the rope twice round

Fig. 29.—Fisherman's Bend.

the spar, in front of the standing part, round the spar again, and then through the last bight.

The "fisherman's bend" (Fig. 29) consists of two round turns round a spar, and a half hitch round

Fig. 30.—Rolling Hitch.

the standing part, and through the turns on the spar, and another half hitch above it, round the standing part. It is used for bending studdingsail halyards to the yard, and, in yachts, for bending on the gaff topsail halliards.

A "rolling hitch" (Fig. 30) is made by taking

Fig. 31.—Topsail Halliard Bend.

three round turns round a spar, and then making two half hitches round the standing part of the rope, and hauling taut.

The "topsail halliard bend" (Fig. 31) is used

Fig. 32.—Racking Hitch.

chiefly on board yachts, and is made by bringing the rope twice round the spar, back round the standing part, under all the turns, over two turns, and under the last. This hitch is shown open for the sake of clearness, but in practice it is usual to jam the coils close together, and haul them all taut.

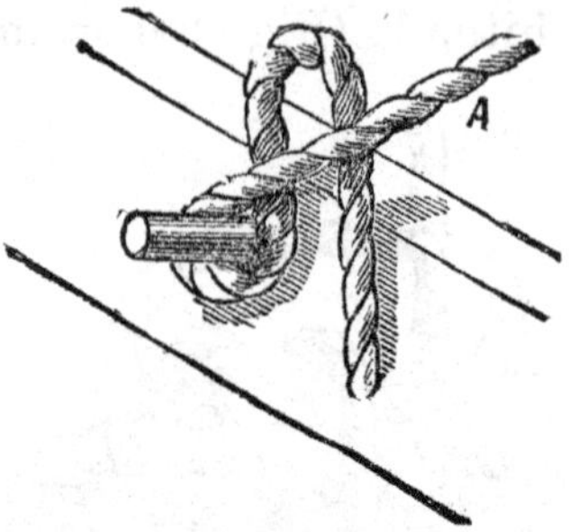

Fig. 33.—Slippery Hitch.

Fig. 32 is a "racking hitch," for hitching a rope on to the hook of a block. Two bights are made in a rope, these are turned over from the

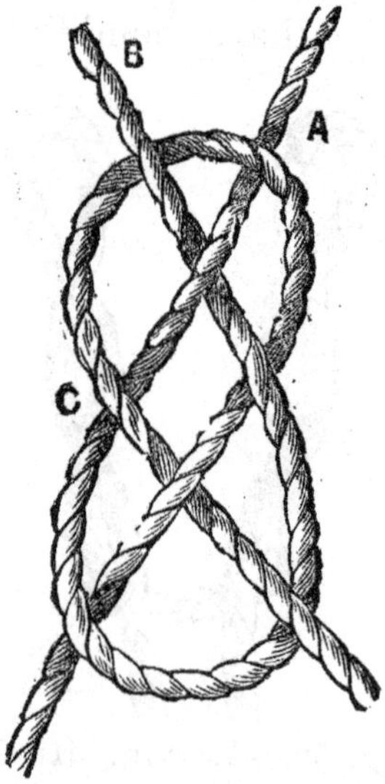

Fig. 34.—Carrick Bend.

operator two or three times, and the two loops are put on to the hook. This is sometimes called a "cat's paw."

The value of the "slippery hitch" (Fig. 33) consists in the readiness with which it can be cast off in case of emergency; at the same time, it holds securely while there is a strain on the rope A. If the mainsheet of small boats is made fast at all, always a more or less risky proceeding, a slippery hitch should always be used as a start. A sharp pull at the end of the rope lets the sheet go at once.

For the "carrick bend" (Fig. 34) lay the end of a rope over the standing part to form a loop; put the end of another rope under the bight over the standing part at A, under the end at B, over

Fig. 35.—Bending Sheet to Clew of Sail.

the rope again at C, under its own part, and over the rope B, and haul taut. The parts A and B form the first loop made. This bend generally is used for binding hawsers together, to increase their length for warping or towing. It can be undone readily without the aid of a pricker or marlinespike, which would have to be used for many knots after they had been in the water. As in the sailor's knot, it is only necessary to grasp the ropes just outside the knot, and push the loops inwards, and the knot comes adrift at once.

Fig. 35 shows the clew of a sail, and the method of bending the sheet on to it. This is termed a "sheet bend." The sheet is not, as many suppose, a part of the sail, but is a rope used in setting a sail, to keep the clew or lower corner of the sail down to its place. In making a sheet bend, the end is passed up through the clew, round the back of it, under its own part, and over the clew again.

Fig. 36.—More Secure Sheet Bend.

The end is generally stopped to the standing part with rope yarn or other small stuff. The knot thus formed is exactly the same as the weaver's knot (Fig. 10, p. 21). Fig. 36 shows a method of giving additional security to this knot. The end is passed twice round the back of the loop before putting it under its own part. This knot is very much used by fishermen in bending a line on to a loop of gut.

Another and somewhat more complicated method of bending a rope on to a loop is illustrated by Fig. 37. B is the standing part, and A the end

Fig. 37.—Bending Rope to Loop.

of the rope to be bent on a loop already formed. Pass the end down through the loop, round over its own part, and through the loop, round the back of it, and through its own bight. When hauled taut, this holds more securely than either of the other methods, but, on the other hand, takes longer to make.

The "Blackwall hitch" (Fig. 38) is a ready way of securing a rope temporarily to a hook. The method of making it is evident from the illustration. As the standing part when hauled upon jams the end against the back of the hook, it holds much more firmly than would be supposed at first sight.

Fig. 38.—Blackwall Hitch.

The "midshipman's hitch" (Fig. 39) is an old-fashioned hitch, used for attaching a tail-block to a rope. A round turn is first made over the standing part, and the end is brought up, passed twice round above the first hitch, and then passed out underneath its own part.

The "marlinespike hitch" (Fig. 40) is used for getting a purchase on the seizing stuff when serving a rope, so as to leave the turns taut. Make a bight in the seizing stuff, and bring it back over the standing part; pass the marlinespike under the standing part, and over the sides of the bight. This is practically identical with the running knot (Fig. 15, p. 24). Seizing is described on p. 96.

Fig. 39.—Midshipman's Hitch.

Fig. 40.—Marlinespike Hitch.

Fig. 41 is a "regulating lashing," used when the tension of a rope requires altering from time to time. Tent ropes are secured this way, as they

Fig. 41 —Regulating Lashing.

Fig. 42.—Stationer's Knot.

require easing in wet weather, and tightening in dry. For this purpose, the piece of wood A is slipped up or down the cord, the friction of the cord against the sides of the hole fixing it sufficiently.

The "stationer's knot" (Fig. 42) is handy for tying up a parcel, as it can be made rapidly, and undone with ease. Make a running noose at the end of a piece of twine, and bring it to the centre of the parcel; take the twine round the parcel again at right angles, round the noose, and making a bight slip it under, as illustrated. A pull at the end releases the knot instantly, as can be proved by experiment.

CHAPTER IV.

RING KNOTS AND ROPE SHORTENINGS.

OFTEN it is necessary to fasten a rope to a ring, and there are a variety of methods of doing this. Fig. 43, for instance, shows the end to be passed through the ring, and a bight put under the standing part; this is a ready way of temporarily fastening the painter of a boat to the ring of a pier; as in the stationer's knot (Fig. 42, p. 41), a sharp pull at the end of the rope frees the painter at once. This is an excellent fastening for many purposes. The "boat knot" (Fig. 44) is another means of mooring a small boat. It is

Fig. 43.—Slippery Ring Knot.

made in the same way as a marlinespike hitch (Fig. 40, p. 40), the only difference being that a

Fig. 44.—Boat Knot.

thowl pin or other small piece of wood is put through the centre of the knot instead of a marlinespike. By withdrawing the pin the knot

Fig. 45.—Lark Boat Knot.

comes adrift of its own accord. Fig. 45 is another form of boat knot, called the "lark boat knot," or "double boat knot." This differs from the last knot, inasmuch as a bight, instead of a single end of rope, is put through the ring; a piece of wood is used to fasten it, as in the boat knot. It is rather the better knot of the two.

If, instead of the ends being brought down outside the bight after it has been passed through

Fig. 46.—Lark's Head.

Fig. 47.—Lark's Head Stoppered.

the ring, they are put through it, a "single lark's head" (Fig. 46) is the result, and Fig. 47 shows the same knot "stoppered." It may be made by passing a bight through the ring, and drawing the two parts of the rope through the bight; or where this is not practicable by reason of one end of the rope being fast, the end may be passed up through the ring behind the standing part, and down through the ring and bight again. Sometimes, instead of stoppering this knot with an overhand

Fig. 48.—Lark's Head with Crossed Ends.

knot, as in Fig. 47, the end is seized to the standing part with twine.

A "lark knot" with crossed ends (Fig. 48) is

Fig. 49.—Double Lark's Head.

made in the manner above described, except that the end comes over instead of through the bight. The ends are often stoppered as in the last knot.

If the standing part is taken in one hand and the end in the other, and drawn apart, it is nothing

Fig. 50.—Treble Lark's Head.

more than a "clove hitch" or "builder's knot" (Fig. 25, p. 31) under a different name.

In the "double lark's head" (Fig. 49) a bight is first made, and the ends passed through it; the ends are then put through the ring and through the loop just made and hauled taut. Fig. 50 shows

Fig. 51.—Backhanded Sailor's Knot.

the "treble lark's head," which is not so difficult as it looks. First bring the bight of a rope up through the ring, take one of the ends, and pass it through the bight, and up through the ring,

Fig. 52.—Capstan Knot.

and then down through its own bight; do the same with the other part and the knot is formed.

In Fig. 51 is illustrated a "backhanded sailor's

Fig. 53.—Another Form of Sailor's Knot.

knot," made by passing an end through the ring round at the back of the standing part and through the ring again, and finishing with two half-hitches round the standing part. It may also be made with the end in the last turn put under the standing part and under its own part.

Fig. 52 is the "capstan knot." To make this, cross the end of the rope after it is through the ring, bring it round the standing part, through the first bight and through its own bight, thus forming a sort of figure-of-eight knot.

Another "sailor's knot," composed of two half-hitches round the standing part of the rope, is shown by Fig. 53. This is one of the most useful

Fig. 54.—Gunner's Knot.

and easily-made knots known, and is used as a mooring knot.

A "gunner's knot" (Fig. 54) is simply a "carrick bend" (Fig. 34, p. 36), mado with the two ends of a rope after it has been passed through two rings. Gunners themselves call this a "delay knot."

The knot shown in Figs. 55 and 56 is called "manharness," or the "artilleryman's knot," and is used when hauling guns over a difficult country, when horses cannot be employed. It is a valuable knot where heavy weights have to be drawn with ropes, as by its use a man can exert his strength

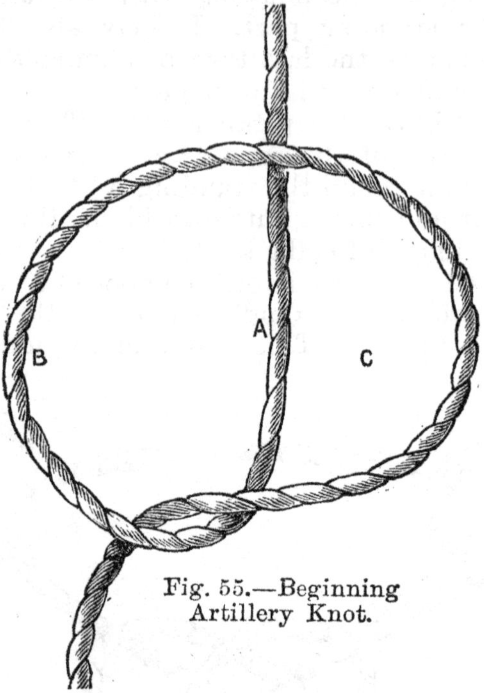

Fig. 55.—Beginning Artillery Knot.

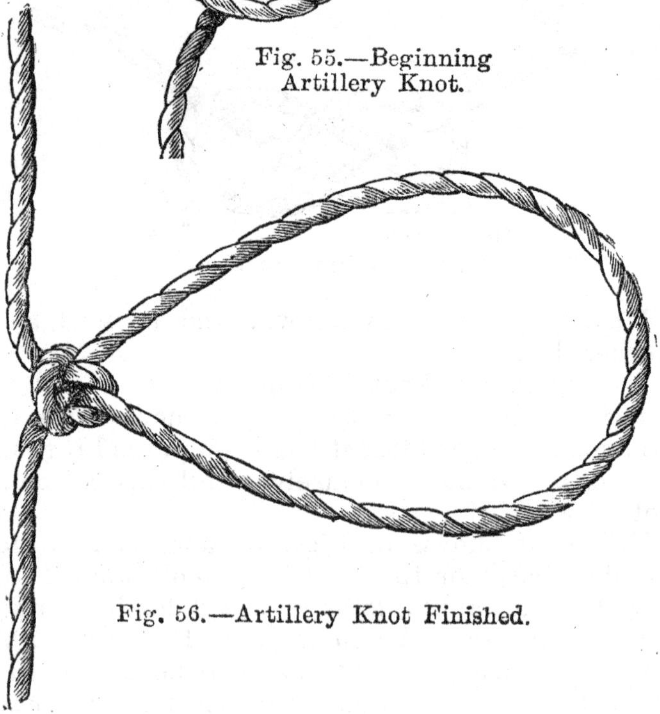

Fig. 56.—Artillery Knot Finished.

to much greater advantage than by merely grasping the rope with his hands. To make it, form a half-hitch, turn it round, and lay it against the standing part. This is, in fact, a marlinespike

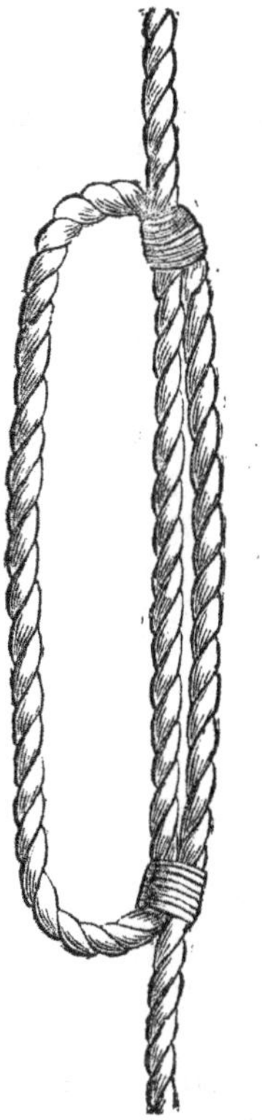

Fig. 57.—Bend Shortening.

Fig. 58.—Chain Knot.

hitch, and is represented in Fig. 40, p. 40. Now pass the right hand into the bight C, and going under A, as shown in Fig. 55, grasp the part B, and draw it through C until a loop of sufficient size is made. When using it, the head and one arm are passed through the loop illustrated in Fig. 56, which shows the knot finished. A little care is required in closing the knot, so that the turns may jam properly one against the other, or the knot will slip. When several men are employed, a bow-line is generally made at the end of the rope, and as many loops as there are men to haul at equal distances along it.

It frequently happens that a rope is too long for its purpose, and as it is inadvisable and wasteful to cut it, some method of reducing its length has to be devised; hence have arisen what are termed "shortenings."

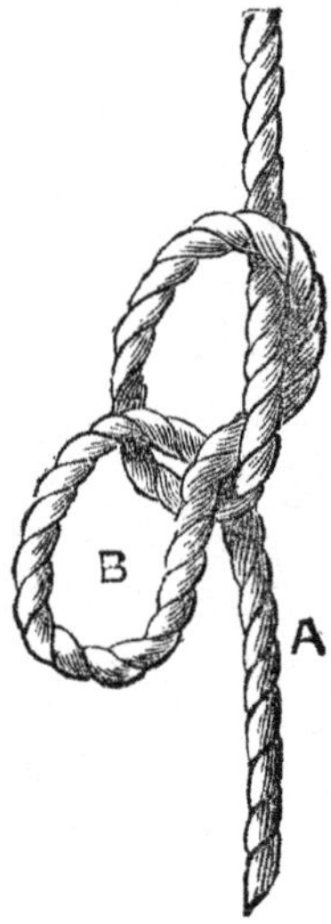

Fig. 59.—Beginning Chain Knot.

The "loop" or "bend" shortening (Fig. 57) is the simplest of these. The rope is merely bent as much as is required, so as to form two bights, and the two parts seized together with small stuff. This is a simple and good method of shortening ropes.

The "chain knot" (Fig. 58) is another method of shortening ropes. In beginning it, make a running knot (Fig. 59), and draw a portion of the part A through the loop B; do this with the forefinger and thumb of the left hand. It will now form another loop, through which a fresh piece of A is to be passed. This process is to be repeated until all the slack of the rope is taken up Finish it off by putting a piece of stick or a belaying-

Fig. 60.—Twist Knot.

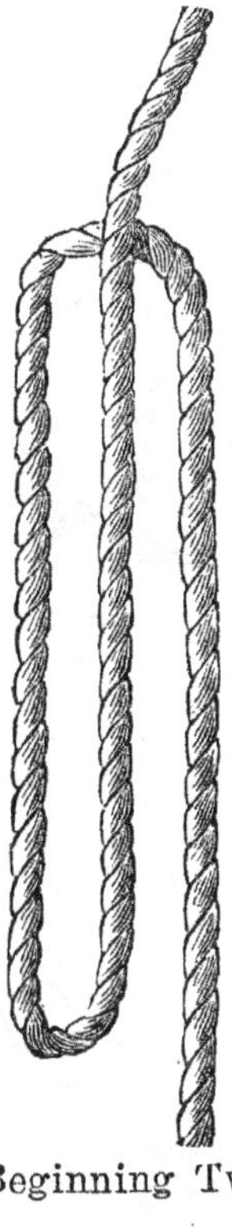

Fig. 61.—Beginning Twist Knot.

pin through the last loop, or by drawing the end itself up through the bight.

The "twist knot" (Fig. 60) is another shortening. Place the cord to be shortened as in Fig. 61, bring one of the outside parts over the middle strand, and the outer strand on the opposite side is brought over this, which is now the middle part. This is continued as long as required, the outside strand alternately being placed over the centre strand, as in an ordinary three-plait. Fig. 60 shows the plait completed, and the manner of

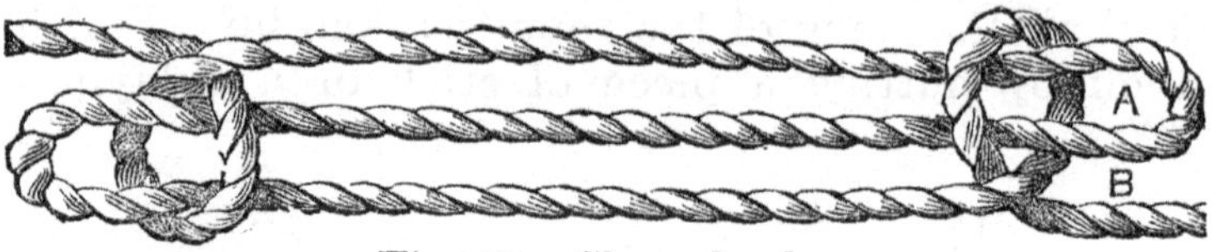

Fig. 62.—Sheepshanks.

finishing off the end. These shortenings are ornamental as well as useful. They may be used for thickening a piece of small cord, so as to give

Fig. 63.—Sheepshanks Fastened.

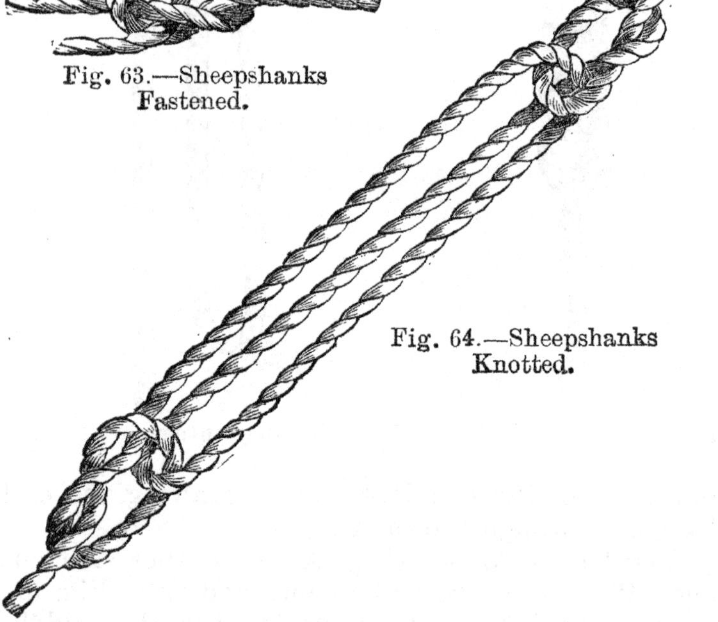

Fig. 64.—Sheepshanks Knotted.

more substance where it has to be hauled upon, and thus prevent the hand being cut with the cord.

Fig. 62 is the "sheepshanks" or "dogshanks"

—a common and old-fashioned method of shortening, applicable to any size of cordage. The twist and chain knots are not suitable for very stout rope. Two bights are formed in the rope, as in Fig. 61; a half hitch is then made at each end, and slipped over the bights. This is made more secure if a seizing is put round the two parts, A, B (Fig. 62). It can also be secured when ends of

Fig. 65.—Sheepshanks Knotted.

rope are free by passing each end through the adjoining bight, as shown in Fig. 63.

Fig. 64 is a "knotted sheepshanks." It differs from the last in that the ends are fastened rather differently. The two bights are made as before, and each end, after passing through the bight nearest to it, is put through the bight it has just made, thus forming an overhand knot at each end. In making Fig. 65, the parts of the rope are arranged as before. A marlinespike hitch (Fig. 40, p. 40) is made at each end and the bight put through it. The left side of Fig. 65 shows the knot

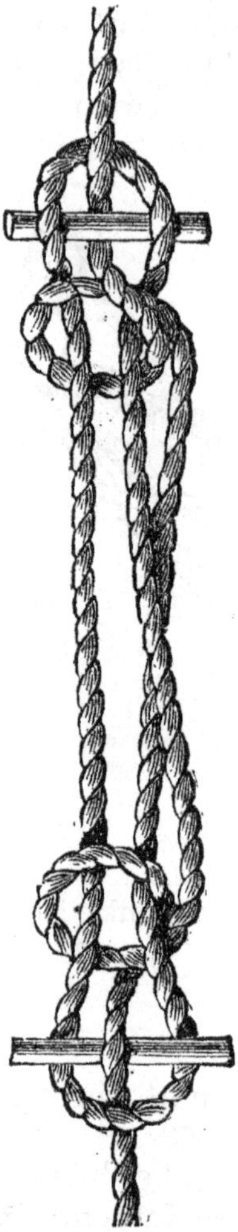

Fig. 66.—Boat Knot Shortening or Sheepshanks Toggled.

Fig. 67.—Knot Shortening.

made, and the bight in the act of being passed through it. It goes over the outside strand, under the centre one, and over the next. On the right side of the figure the loop is shown in its place, ready to be hauled taut.

The "boat-knot" shortening (Fig. 66) is another form of sheepshanks. The ends are secured by

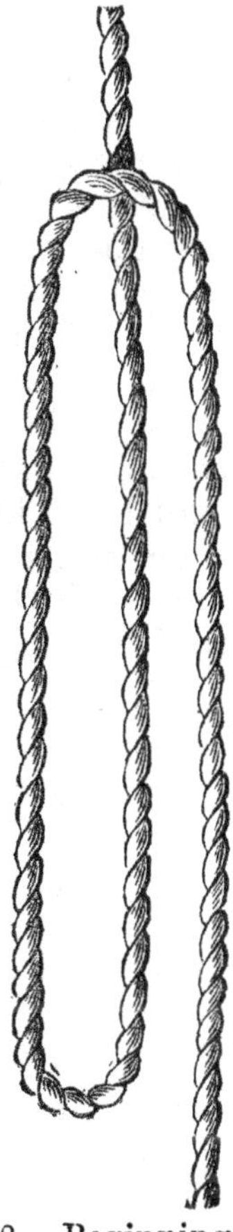

Fig. 68.—Beginning Knot Shortening.

Fig. 69.—Double Chain Knot.

bringing a portion of the loose part of the rope through the bight at each end, and toggling it with a belaying-pin or piece of wood, as in the boat knot (Fig. 44, p. 44). It is loosened readily, and can be made when both ends are fast.

The "knot" shortening (Fig. 67) is a ready mode of taking up the slack of a rope, though it is not suitable for very stout ropes, and can be made only where one end of the rope is free. The rope is laid as at Fig. 68. The three parts are grasped with both hands near the bights, and an overhand knot is formed with the whole of the strands. It forms a good shortening for moderate-sized cordage, where the strain is not too heavy.

The "double chain knot" (Fig. 69) is perhaps the most ornamental knot of this group. A turn is first taken round the standing part, and the loose end is then passed through the loop thus formed at A. In doing this another loop B is made, through which the end is brought. The end is thus continually passed from one side to the other through the preceding loop until the knot is of the proper length. It may be finished, if desired, by making an overhand knot with the end over its own part, or merely passing it through the last loop, and hauling on it.

CHAPTER V.

TIES AND LASHINGS.

A "WEDDING knot" or tie, used for fastening together the eyes at the ends of two ropes, is shown by Fig. 70. It is made by passing rope-yarn or marline through the eyes backwards and forwards

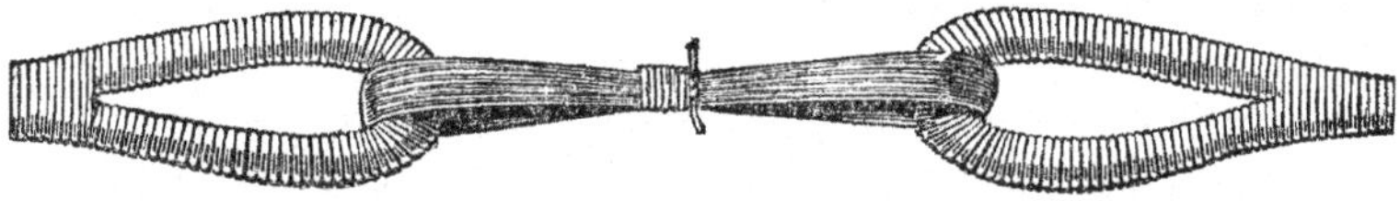

Fig. 70.—Wedding Knot.

until strong enough, and then is fastened by taking several turns round the middle and fastening the ends with a reef knot. This forms a sort of hinge between the ropes.

The "chain knot," for lashing to a spar is illustrated by Fig. 71; a clove-hitch is first formed round the spar, and as many single hitches as required are then made. It may be finished off with any secure knot. Fig. 72 shows another and better way of making the chain knot. An overhand knot

Fig. 71.—Chain Knot Lashed to Spar.

is formed at each turn, and consequently it is much more secure than Fig. 71. This is used for bending yachts' sails to the gaff. As each turn forms a knot if the cord parts, the remainder holds firm, and does not necessarily come adrift, as it would be almost sure to do it if fastened as in Fig. 71.

Fig. 73 is a "cross lashing," employed when a lever is used to a rope. After several turns round the rope, the lashing is crossed round the

Fig. 72.—Improved Chain Knot.

lever and fastened with a reef knot. All these lashings are used when several men are required to haul on large ropes at the same time.

For the "necklace tie" (Fig. 74) several turns are taken round the spar to be joined, then two turns round the lashings, and it is secured with a reef knot. When this is used as a lashing for shearlegs, the crossing of the two legs puts a strain on the knot, and effectually secures it. For this purpose it is called a Portuguese knot.

Fig. 75 shows a "packing knot," used for securing large pieces of timber together. It is used near stone quarries for holding the blocks of stone on to the carriages by which they are moved. Fig. 75 represents a block of granite secured to a trolley with packing knots. Two or

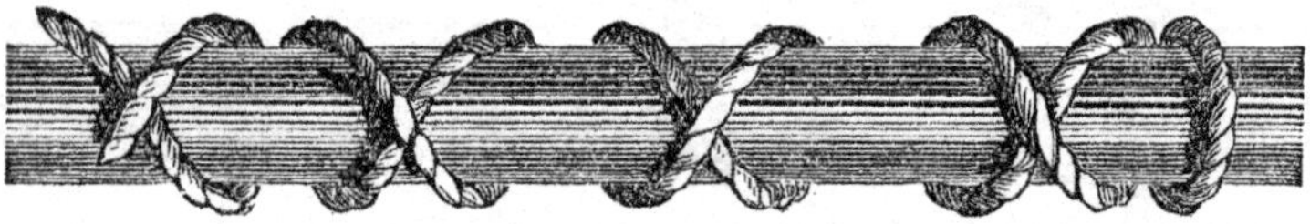

Fig. 73.—Cross Lashing.

three turns are made somewhat loosely with cordage round the block and its carriage ; a stout piece of wood is then inserted under the coils, and twisted round until all the slack is taken out and the cordage is taut. The end of the lever is then secured with twine to the side of the carriage, as shown in the right side of the figure. The other lashing is supposed to be all ready for tautening up.

It is often necessary to lash two things together

Fig. 74.—Necklace Tie.

without showing an external knot, which would spoil the smoothness and neatness of the work—as, for instance, in whipping the two parts of a broken fishing-rod together. Fig. 76 shows a common method of finishing off whipping without showing a knot. Lay one end forward, as at A, then pass the other end round and round a sufficient number of times, hauling taut each time; three or four loose turns are now made, and

Fig. 75.—Packing Knot.

the end passed under them backwards; these are worked down into their places, and when the ends are hauled taut and cut off the job is completed. The end A need not come so far as shown in Fig. 76, but may be hidden under the coils.

Fig. 77 is another method of accomplishing the purpose. Instead of a single end, as in the last case, a bight of the seizing stuff is laid along the part to be whipped, and the turns passed over it; when these are completed the end is passed

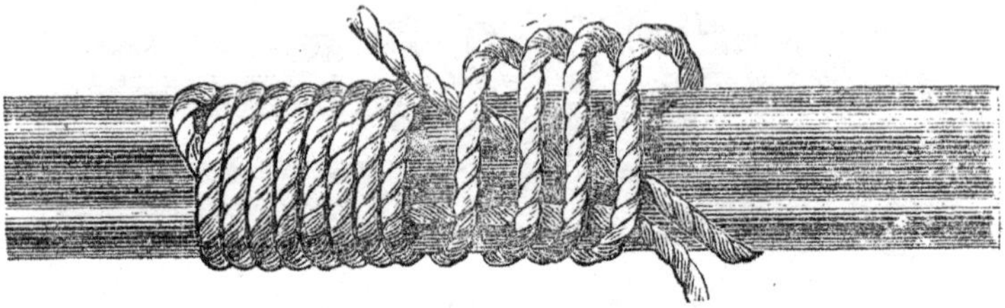

Fig. 76.—Finishing off Whipping.

through the bight, as at A. The end B is now hauled upon to bring the bight and the end of the rope snug under the coils. There are now two loops interlacing at the centre of the work, and these cannot come undone. When the ends A and B are cut off close to the turns, the whole is fair and smooth.

"Nippering," or "packing," is shown in Fig. 78. This is a method of securing two ropes together with cross turns; these are hauled taut,

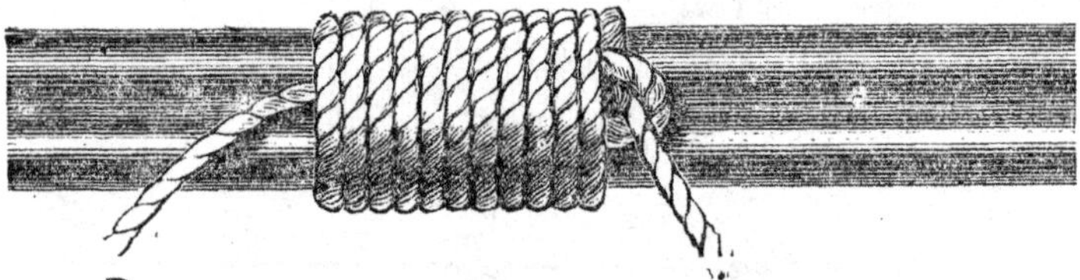

Fig. 77.—Finishing off Whipping.

jambing the ropes together, and are further secured by round turns over all, with a reef knot at the ends.

The "west country whipping" (Fig. 79) is an excellent method, and deserves to be practised oftener than it is. Bring the middle of the material used under the part to be whipped, raise the ends and tie an overhand knot, lower the ends and tie another underneath; continue tying a single knot above and below alternately, finish-

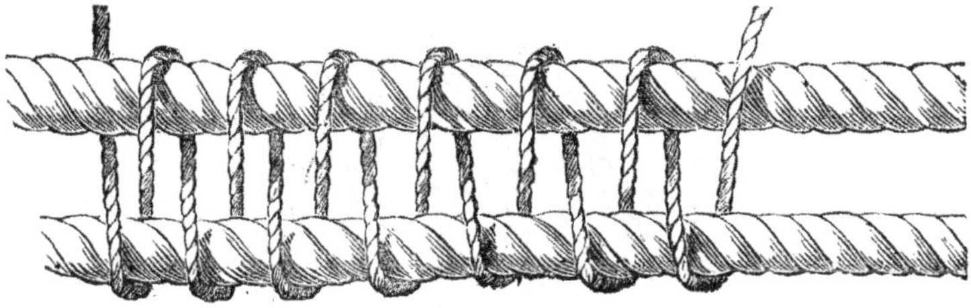

Fig. 78.—Nippering.

ing with a reef knot; or a round turn or two may be taken and the ends may then be secured; but a reef knot is the most usual way of fastening off this whipping. This is not quite so neat-looking a method as Figs. 76 and 77, but it is very strong and trustworthy, and is an excellent way of fastening large hooks, such as those used for cod or conger, on to a line.

A "catspaw" (Fig. 80) is used for attaching a

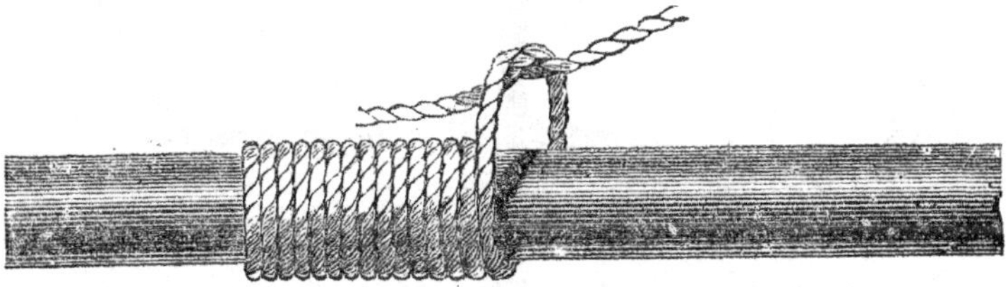

Fig. 79.—West Country Whipping.

rope to a tackle hook. Fig. 81 shows how to begin it. A loop is made, and laid over the standing part so as to form two bights; these are rolled over two or three times, and the hook inserted in them. When the standing part is hauled upon, the

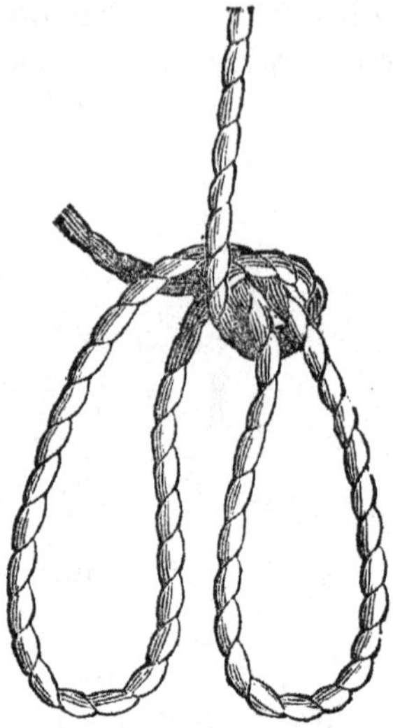

Fig. 80.—Catspaw.

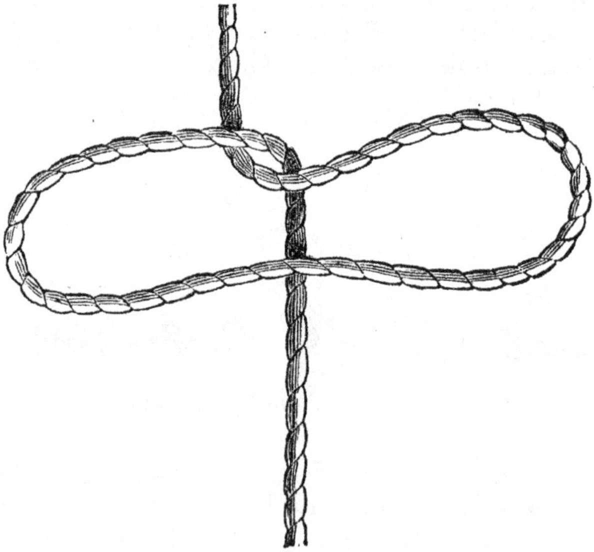

Fig. 81.—Beginning Catspaw.

hooks take the form shown in Fig. 71 (p. 59), and will not slip.

Fig. 82 shows a way of securing a block to a

Fig. 82.—Securing Block to Rope.

rope with a selvage strop. The middle of the selvage is placed against the rope, and cross turns taken until the bights come together, when the loop of the block is put through them.

CHAPTER VI.

FANCY KNOTS.

ALTHOUGH these knots are termed fancy knots, they are not necessarily used for ornamental purposes, but are often of considerable utility; indeed, they could hardly be done without aboard ship.

One of the commonest knots of this kind is the "wall knot"; Fig. 83 shows this ready for haul-

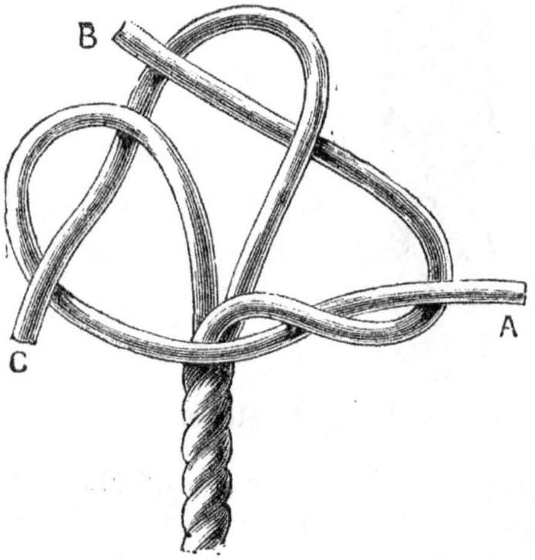

Fig. 83.—Wall Knot.

Fig. 84.—Stopper Knot.

ing taut. First unlay the strands at the end of a rope and make a bight with one strand A; hold this to the standing part with the thumb of the left hand, make a loop with the next strand B round the end of the first strand, and bring the remaining strand C round the end of the strand B and through the bight of A.

If the ends are taken round once more and

brought up in the centre of the knot, it is called a "stopper knot" (see Fig. 84). In this case the

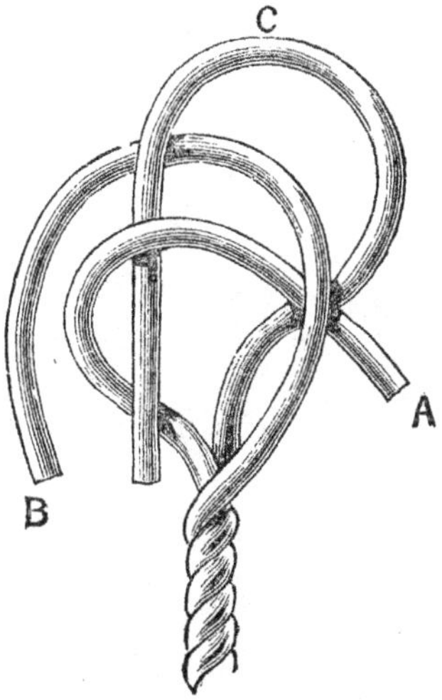

Fig. 85.—Beginning Crowning.

ends are whipped together and cut off level. Fig. 84 is the stopper knot finished. Fig. 85 shows the crowning begun. Open the strands of a rope as before, but do not put a seizing round them.

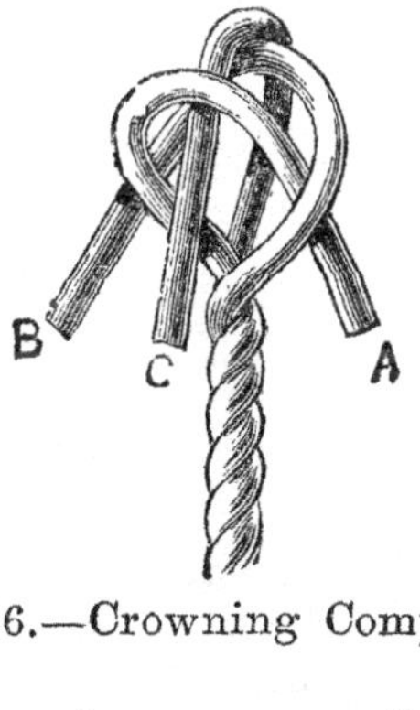

Fig. 86.—Crowning Complete.

Lay the strand A down over the centre of the rope, and then bring B down over A and strand

C over B and through the bight of A. Fig. 86 shows how the strands tie when they are nearly taut. The strands in Fig. 85 are hardly in the position which they occupy when the knot is actually being made, as they are then much snugger.

Fig 87.—Beginning Manrope Knot.

Crowning is used by itself as a method of preventing the strands of a rope unlaying while in use. In this case, after crowning as above, pass one end over the next strand in the standing part, and under the following one. Do the same

Fig. 88.—Manrope Knot.

Fig. 89.—Tack Knot.

to each of the other strands in succession. This may be repeated and the ends cut off. Masons, whose ropes have to stand a good deal of knocking about, generally use this plan; for this reason it is called "masons' whipping." Though very strong and standing hard usage well, this is not the neatest way of finishing the ends of a rope.

Crowning may also be used in connection with other knots. For instance, it is possible to crown first and wall afterwards, as shown in Figs. 87 and 88. Fig. 87 shows the crowning in the centre of the knot hauled taut; this is made on the end of a rope as just described. Now make a single

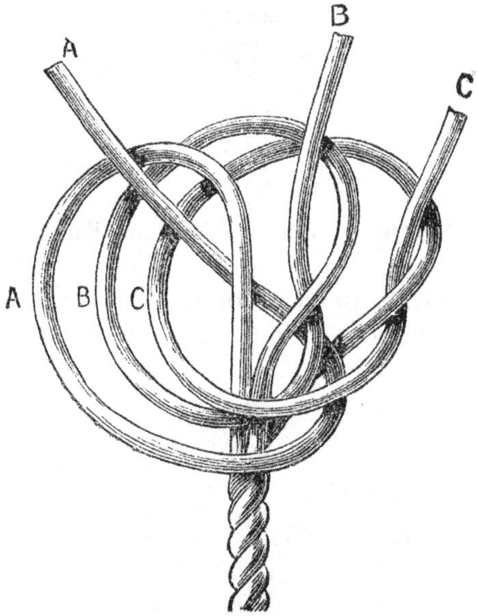

Fig. 90.—Beginning Matthew Walker Knot.

wall knot under the crowning. When the knot is tightened it will appear as in Fig. 88, and is called a "manrope knot." A single "wall knot" may be crowned after it is made; a single wall and a single crown are the result. Wall again by passing one end under the part of the first walling next to it and bring it up through the same bight, and do the same with the other two strands, and the result is a knot with a double wall and a single crown. A double-walled double-crowned, called a

"Tack Knot," is made as the last knot—that is, double-walled and single-crowned. Now lay the ends by the sides of those in the single crown, and

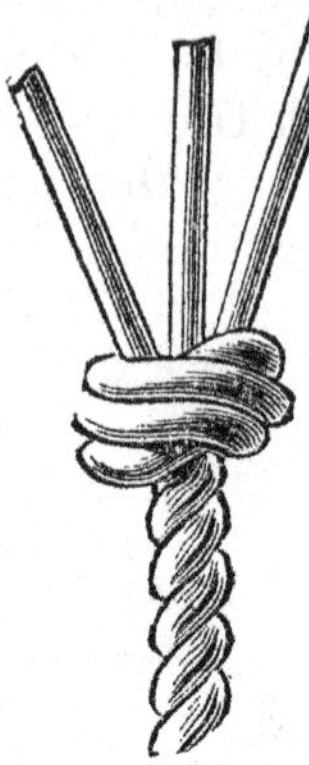

Fig. 91.—Matthew Walker **Knot.**

with the aid of a pricker bring them down through the double walling and they will be alongside the

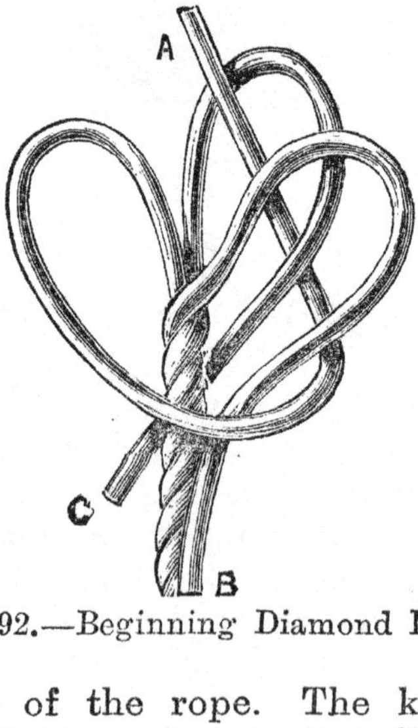

Fig. 92.—Beginning Diamond Knot.

standing part of the rope. The knot is shown completed, with the ends cut off, in Fig. 89.

In Fig. 90 is shown a Matthew Walker knot

open ready for being hauled taut. After putting a seizing round the rope and unlaying the end as before, bring one strand A round the rope and put it through its own bight, the next strand B underneath, through the bight of A and through its own bight, and the last strand C underneath through both the other bights, and lastly through its own bight. Fig. 91 shows the knot completed.

The "diamond knot" is an ornamental knot made some distance from the end of a rope; it is

Fig. 93.—Diamond Knot before Hauling Taut.

therefore necessary to unlay the rope considerably more than is required for the preceding knots. To form a diamond, bring each of the three strands down alongside the standing part of the rope, thus forming three bights, and hold them thus with the left hand. Take the first strand A (Fig. 92), and, putting it over the next B, bring it up through the bight of the third strand C. Take the end of the second strand over the third and up through the bight of the first. The last strand is brought over the first and up through the bight of the second. Haul taut and lay the rope up again. Fig. 92 is the way the knot is begun, showing the manner of taking the first strand. Fig. 93 shows the loops in their places with the ends through them before they are hauled taut, and Fig. 94 shows the knot finished. Remember that, after the bights are formed down the standing part, each end successively goes over the strand next to it and up through the loop beyond. This knot is the "single diamond."

For a "double diamond" (Fig. 95) make a single diamond as above, without laying up the strands;

Fig. 94.—Single Diamond Knot.

Fig. 95.—Double Diamond Knot.

the ends are then made to follow the lead of the single knot through two single bights, the ends coming out on the top of the knot. The last strand passes through two double bights. When the ends are hauled taut they are laid up as before. The last four knots are used for the ends of lanyards, man and ridge ropes, yoke lines, etc.

The "shroud knot" (Fig. 96) is of use in joining two ropes together, particularly for joining a stay or shroud that has been carried away. The ends of each rope are unlaid, and placed within one another as in splicing, the parts not unlaid being brought closely together. Make a wall knot

(Fig. 83, p. 66) with the strands of one rope round the standing part of the other rope; turn the ropes over, and do the same with the other set of ends, and they will appear as in the figure. Open the strands, and taper and serve them over if the job is to be particularly neat. Two ropes of different sizes may be twisted in this way, and will be quite secure.

In making the "French shroud knot" unlay the ends and place the two ropes with the strands

Fig. 96.—Shroud Knot.

intermixed as before; bring one set of ends back on their own rope, and make a single wall knot with the other set of strands round the bights of the first set and the standing part. They can then be tapered and served as in an ordinary shroud knot.

For a "spritsail sheet knot," unlay the two ends of a rope and bring the two sets of strands together side by side; these have to be walled together as for a common wall knot. A bight is made with the first strand, the second is put over the first, the third over the second, the fourth over the third, the fifth over the fourth, the sixth over the fifth and through the bight of the first; they are then hauled taut. Crown it by laying two of the strands along the top of the knot and passing the other strands alternately over and under these two, and afterwards hauling them taut. It may be double-walled after crowning by putting the strands successively under the bights on the left of them and through the same bights, and the ends will then come up in the right position to be crowned again. This is done by following the

Fig. 97.—Beginning Turk's Head.

lead of the first crowning and putting the ends through the walling as before.

The "buoy rope knot" can be made on a cable-laid rope only. Unlay the main strands, and take out one of the smaller strands, of which they are composed, from each of the large strands, and then lay them up again. The small strands that have been taken out are now single and double-walled round the rope, and then laid along the divisions after the manner of weaving, and their thin ends stopped with spun-yarn. A stop should be put round the rope with the spun-yarn where the knot is to be made before it is begun, and the walling should be right-handed.

The "Turk's Head" is a highly ornamental knot which, instead of being made out of the rope itself, is formed on the rope with a piece of small stuff. A clove hitch (p. 32) is first made on the rope (Fig. 97); this must be slack enough to allow of the extra strands which will form part of it. Put part A over strand B, thus twisting the two strands; pass the end C under and up through the bight that B now forms, then twist again by

Fig. 98.—Turk's Head.

putting B over A and run the end under and up through the bight of A. Continue twisting the strands by alternately putting one over the other, and at each twist bring the end under and up through the bight which is underneath, A at the beginning going over B; the bight which B makes will be the under one, and therefore the one through which the end C must be passed. The end C must be much longer than illustrated as the whole knot is made with this part, and as the knot when finished contains three groups of three strands each, it is obvious that the length of cord used must be more than nine times the circumference of the rope round which the knot is made. Having made a sufficient number of twists (the exact number depends on the size of the knot), lay the end C alongside D, where it comes out of the knot, and continue following its lead through all its turns as it goes through the knot until the

Fig. 99.—Single Pitcher Knot.

beginning is reached again. There will now be a Turk's head of two parts. If the end is again passed through by the side of the same strand as before, a complete Turk's head of three parts will be formed. Care must be taken to keep the working strand close to and on the same side of the strand that is being followed, or a perfect knot cannot be formed. The first time round is the most difficult, the second is easy enough. Of course, the knot may consist of more parts if required, but three is the usual number. The ends do not require fastening in any way, as in the last round they finish in the middle of the knot under the coils, and are quite secure (see Fig. 98).

The "single pitcher" knot, known also as "Tom Fool's" knot, is shown by Fig. 99. Form two half hitches, as in Fig. 100, one lying halfway over the other. With the finger and thumb of

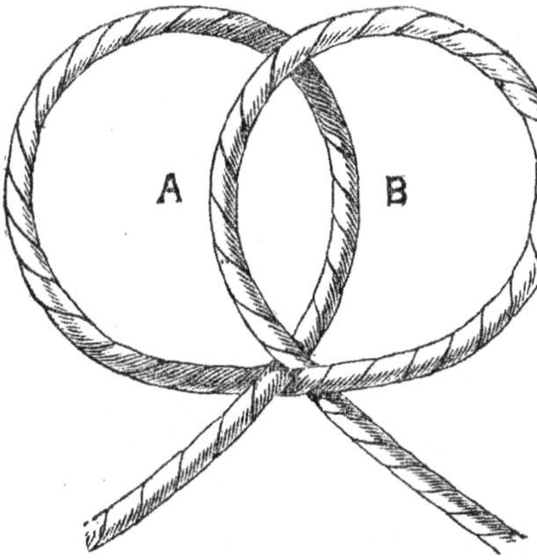

Fig. 100.—Beginning Single Pitcher Knot.

the left hand draw the part A down through the bight, and with the same fingers of the right hand bring the strand B upwards through the bight, under which it lies. Pull out the loops thus formed to a sufficient length and knot the ends together. When used to supply the place of a broken pitcher handle, the centre knot should be hauled taut, and, the pitcher being placed on it, the loops are brought up to form handles. To keep them in their places a lashing is put round the neck of the pitcher, as shown in Fig. 101. This knot is also very useful in slinging a shot when required as a weight, or for any other

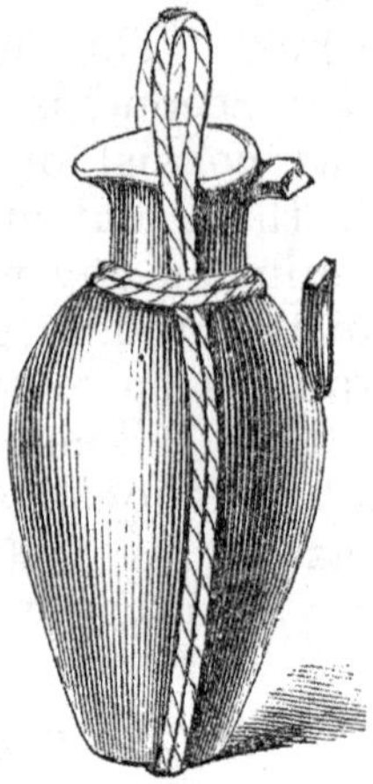

Fig. 101.—Pitcher with Rope Handle.

Fig. 102.—Double Pitcher Knot.

purpose. In this case the centre knot is not hauled taut but left open, forming a large loop on which the shot lies. If the ends are spliced instead of knotted a three-loop knot is made.

This knot is used also as a trick or puzzle knot, and from this arose its name of "Tom Fool's" knot.

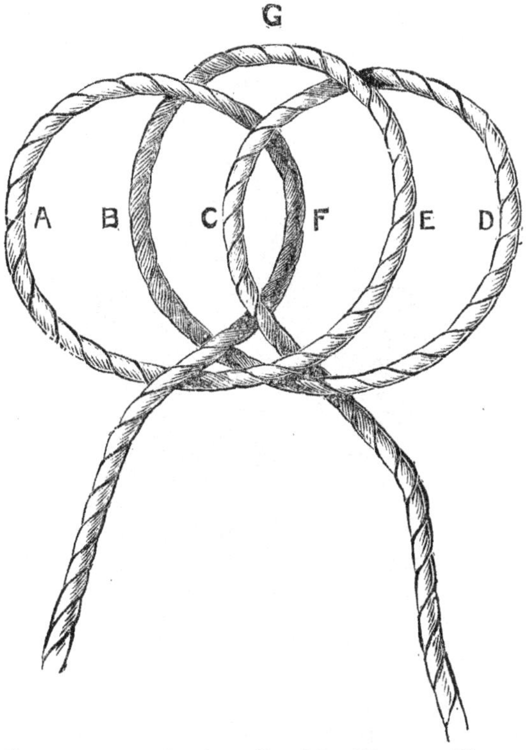

Fig. 103.—Beginning Double Pitcher Knot.

Fig. 102 is the "double pitcher" knot, which goes also by the names of "jury" knot and "true lovers'" knot. It is used as the single pitcher knot to sling a broken pitcher, but in this case there are four loops by which to carry it instead of two. In rigging a jury-mast the end of it is put through the centre of the knot before it is hauled taut; the stays to support and steady the mast are then made fast to the bights of the knot. Form two half hitches in a piece of cord, as in Fig. 103, then make another hitch, which draw behind the other hitches with the inner edge overlapping the inner edge of the first hitch, as shown in Fig. 103. Pass the forefinger and thumb of the left hand over strand A under B and take hold of C.

Fig. 104.—Can or Jar in Sling.

Put the same fingers of the right hand under D over E and take hold of F. Take G between the teeth and draw the three loops out. It is better to make G the length required at first, as the other loops being immediately connected with the ends can be more readily adjusted as to size than the upper loop. When the loops are made the right size the loose ends are spliced together with a short splice, thus forming the fourth loop.

Fig. 104 shows a ready way of slinging a can

to improvise a paint pot, to dip for water, etc. Pass the end of the cord under the bottom of the can and bring the two parts over it, and make with them an overhand knot; open the knot, as shown in Fig. 105, and draw the two parts down until they come round the upper edge of the can; haul taut, and knot them together again over the can, as shown in Fig. 104.

If the ends of the shamrock knot (Fig. 106) were spliced, a four-looped knot would be formed. Though used for the same purposes as the double pitcher knot, it is not so good as that, being more troublesome to make and not so strong, in con-

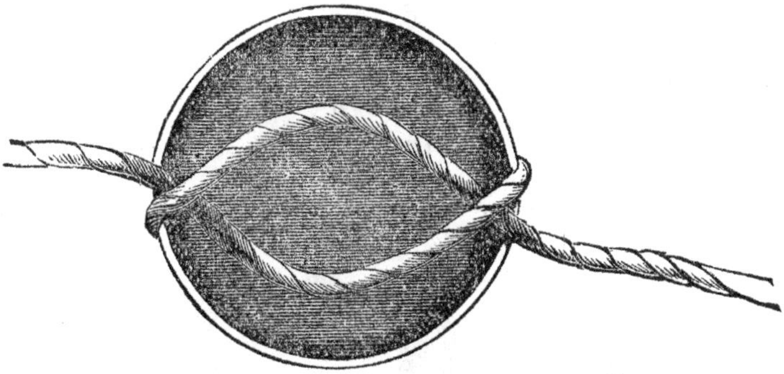

Fig. 105.—Beginning Can Sling.

sequence of the short nip of the strands in the centre of the knot. Fig. 107 shows the way of making it. An overhand knot is first formed with the ends at A; the end B is then laid across the upper loop, brought round and under the right loop and up through the bight C. The strand D, after passing at the back of the upper loop, is carried over the left loop and down through the bight E. The loops are now adjusted for size and the knot hauled taut. Fig. 108 gives another way of making this knot. Two overhand knots intersecting one another are made on the ends, as illustrated; the part A is drawn up through the

Fig. 106.—Shamrock Knot.

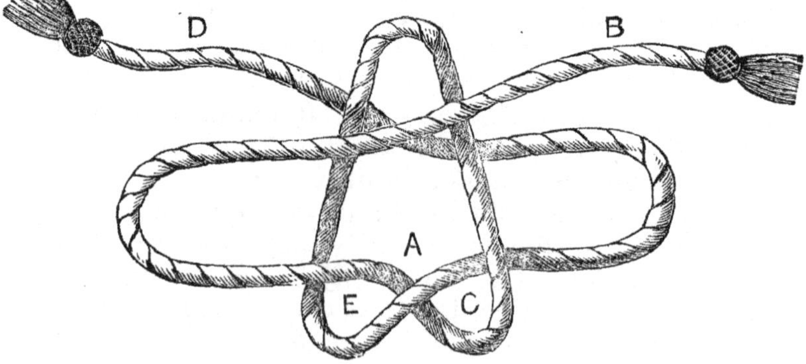

Fig. 107.—Beginning Shamrock Knot.

bight C, and the part B down through the bight D. These form the side loops, and the top loop being pulled out the knot is completed. By an extension of these methods knots may be made with any number of loops, but the difficulty increases greatly as the loops increase, so much so, that

Fig. 108.—Another Way of Making Shamrock Knot.

many loops cannot be made except wire replaces the cord.

The dalliance knot whose beginning is shown by Fig. 109 is a trick knot difficult to learn when it is merely seen rapidly made. The object is to make two independent double-knots at once on a double cord. Double the cord so that the ends lie together; bring the bight over the standing parts, as shown in Fig. 109, and cross the strand A over the strand B; they will now appear as in Fig. 110. Press the part C down between the two strands on which it lies, and bring it up through the opening D, draw it out, and two overhand knots will be formed on the double cord. While

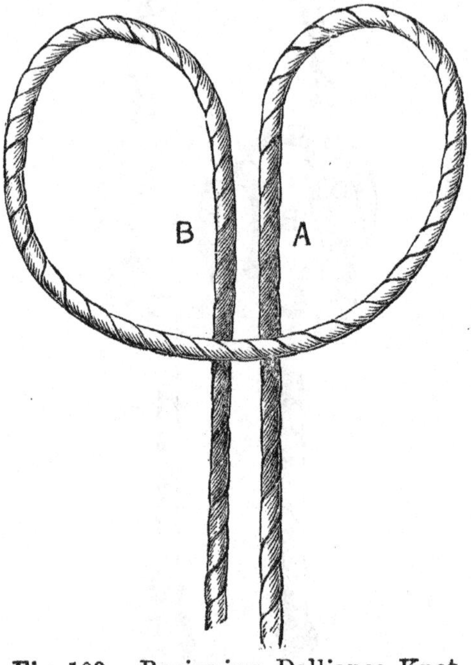

Fig. 109.—Beginning Dalliance Knot.

the part C is being drawn out through D, the whole of the loop E must be brought up through the bight F; this forms the upper knot. The lower knot is made by loop F, C forming bight at top of double cord. The finished knot is practically the same as Fig. 16 (p. 25).

Some stage performances many years ago consisted of various rope tricks. In the principal one the performers were shut up in a cabinet, and when the doors were thrown open they were found seated on two chairs tightly bound hand and foot. After examining the knots, the doors were closed and the men rang bells, played on the tambourine, and threw things out of a small window in the top of the cabinet. On the doors being opened again directly they were found firmly tied to their chairs as before. They were able to do this by means of an ingenious knot,

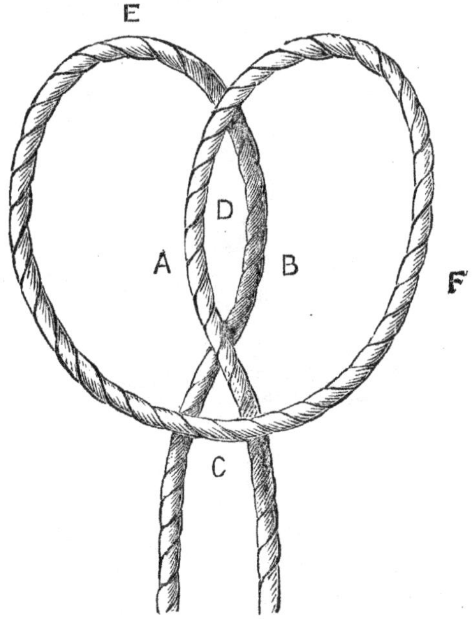

Fig. 110.—Next Stage in Dalliance Knot.

shown in Fig. 111. To perform the trick, two medium ropes each about 12 ft. long are required. First the openhand knot A (see Fig. 8, p. 19) joining the two ropes is made, the ends being passed twice through the bight to increase the size of the knot. Two running knots are made close up to this knot as at B, B. The knotted end of the rope is laid on the seat of a chair with the ropes passing down the back of the seat and under the chair. The performer seats himself on the chair, and, drawing the loose ends of the ropes up in front from under it, he passes them round and round his legs and the legs of the chair in a complicated manner. He draws the knotted end from under him, and, putting his arms over the back of the chair, passes his left hand down through one loop and his right hand up through the other. He now turns his right hand down until the palms of both hands are together and the fingers pointing down-

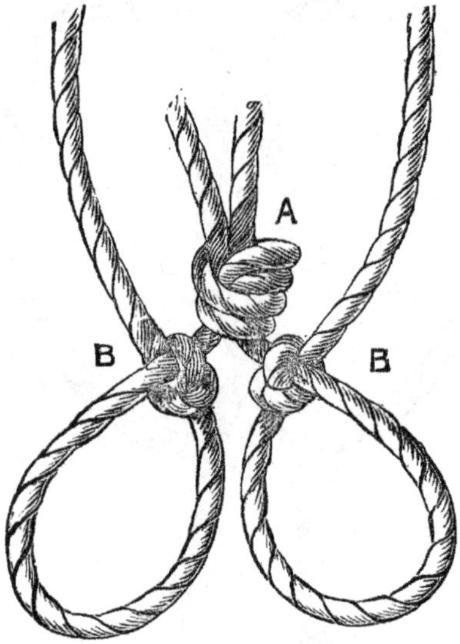

Fig. 111.—Davenport Brothers' Knot.

wards. This produces a twist in the ropes which takes up the slack and tightens the cord round the wrists. The large knot being between the hands effectually hides this, and the wrists merely appear to be as tightly bound together as they can be. The performer has merely to reverse this last proceeding—that is, to bring the right hand up again, and so undo the twist—and his hand can be withdrawn as readily as it was put into the loop. The trick requires some practice, and the

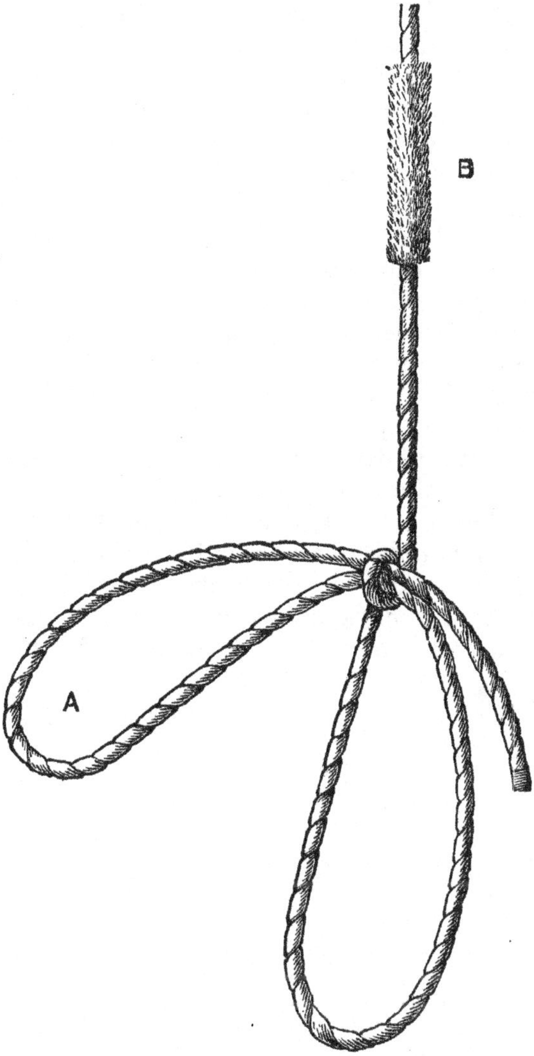

Fig. 112.—Bell-ringer's Knot.

size of the loops must be regulated by the size of the performer's wrists. The knots should also be so placed on the chair at the commencement that the ropes are tight when the hands are in the loops.

Fig. 112 is the "bell-ringer's knot," but really it is a hitch and not a knot. Church bells have a large wheel on the axle on which they are hung, round which the bell-rope passes; this is done to obtain sufficient leverage to raise the bell mouth upwards when it is rung. This requires a long rope, a good portion of which lies on the belfry floor when the bell is down. When the ringing is over this slack is hitched up out of the way in the manner shown. The loop A is made near the end of the rope, laid against the standing part, and a hitch taken over it at about the height of a man's head. The hitch should be kept quite close to the standing part, and it will hold the loop quite securely; at the same time a slight pull at the end releases the whole thing at once. The part B where the rope is grasped when the bell is checked as it comes over is called the sally or tufting. It is made by opening the strands and inserting short pieces of worsted, which are afterwards trimmed until they are all of one length.

CHAPTER VII.

ROPE SPLICING.

Splicing is a method of joining ropes by interweaving together their strands. When ropes are to run through blocks they cannot be joined by knotting, as the knot would prevent their passing through the block. In this case they are always united by splicing. In driving ropes also knots are out of the question.

For the short splice, the ends of the two ropes are unlaid for a sufficient distance, and placed

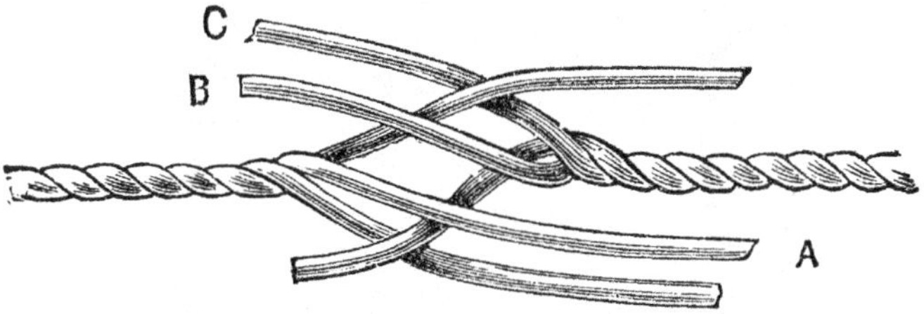

Fig. 113.—Beginning Short Splice.

together, as in Fig. 113, the strands of one rope going alternately between the strands of the other. The two ropes are then jammed closely together. The end of one rope with the strands of the other rope is now held firmly in the left hand. Sometimes a lashing is put round the strands to keep them down to the rope on which they lie. Pass the middle strand A over the strand of the other rope B which goes down to the left of it; then bring it under C, and haul taut. Do the same to each of the other strands in succession, putting them over the next strand to them and under the next beyond. Turn the rope round and do the same to the other set of strands; this may be repeated on both ropes. Care must be taken not to bring two strands up through one interval in the rope. Each strand should come up separately between two strands of the rope they are passed

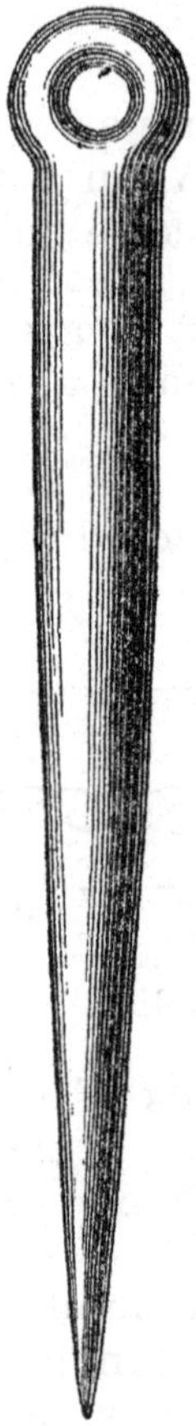

Fig. 114.—Marlinespike.

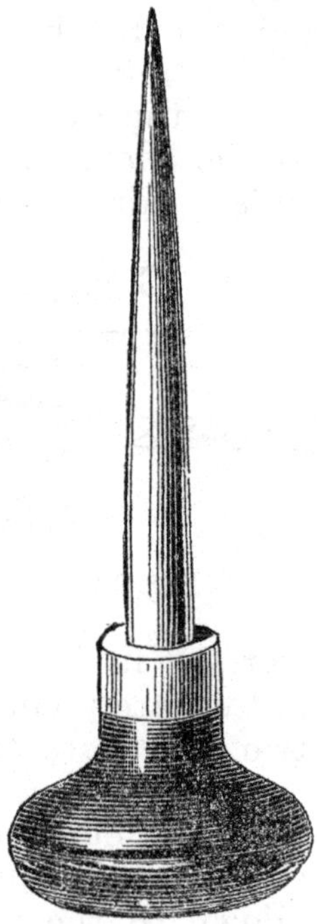

Fig. 115.—Pricker.

into. If it is desired to taper the splice for the purpose of making it neater after the ends have been interwoven, divide the yarns of which the strands are composed, pass one-half as before, and cut off the other half. To bring the rope to shape again after splicing, roll it under the foot; if small cord has been used, a piece of flat wood is substituted for the foot.

As the strands of a rope are tightly twisted together it requires some force to open a passage for the parts of one rope through the other. For this purpose, in the case of large ropes, is used a marlinespike (Fig. 114) made of iron, copper, or hard wood. Copper is preferable, as it does not rust like iron or break like wood. With small stuff a steel pricker (Fig. 115) is used. A fid is employed for very large ropes, this being merely

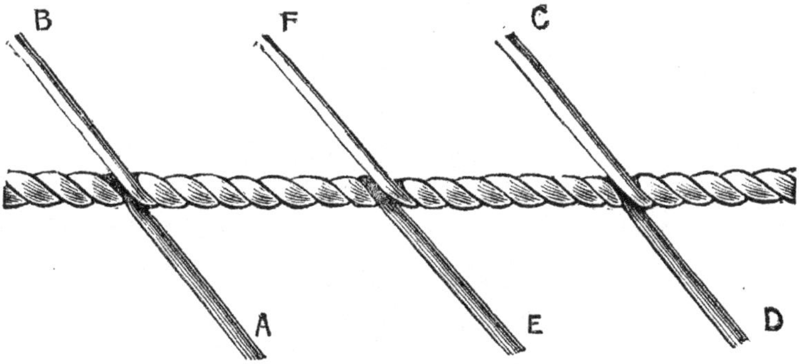

Fig. 116.—Long Rope Splice.

a tapered wooden pin generally made of lignum vitæ.

The long splice illustrated by Fig. 116 has advantages over a short one. To make it, unlay the ends of two ropes for a much greater distance than before and put the ends together. Unlay one strand A for some length, and fill up the space left by its removal with the opposite strand from the other rope, as B. Do the same with two more strands C D; C is the one unlaid, and D the one laid up in its place. Make an overhand knot with the two remaining strands E F, taking care that the ends follow the lay of the rope and not across them. Divide both strands into halves and pass one-half over the next strand, and under the following one; do this two or three times and cut all the ends off close. Work the remaining two pairs of strands the same way and the splice is finished. The rope should be well stretched before the ends of the strands are cut off.

Fig. 117 is a cut splice forming an oblong loop in the middle of a rope. The end of one rope is spliced into the standing part of another, as at A, A (Fig. 118), so as to form an eye. The end of the other rope is then spliced into the standing part of the first rope, and the spliced parts served over.

A neat way of forming an eye at the end of a rope (it is known as the eye splice, Fig. 119) is by unlaying the strands and placing them on the standing part so as to form an eye, then put one

Fig. 117.—Cut Splice.

strand under the strand next to it, and pass the next over this strand and under the second; the last strand must go through the third strand on the other side of the rope. Taper them as before by halving the strands and sticking them again.

To make a cable splice unlay the ends of the ropes to be joined for some distance, place them together and make a short splice. Leave a suitable length, and thence reduce each strand to a long taper by gradually cutting away as many yarns as necessary; neatly point over the taper and lay the ends in the intervals of the rope. Put a seizing at each end of the splice, an end seizing at the beginning of the pointing and a stop at the end of the tails. This is the best splice for cables, as it may readily be undone.

Another method of making a cable splice is to splice the ends in twice each way, then to pick out the strands, worm part of them round the cable, and taper away the rest, which should be

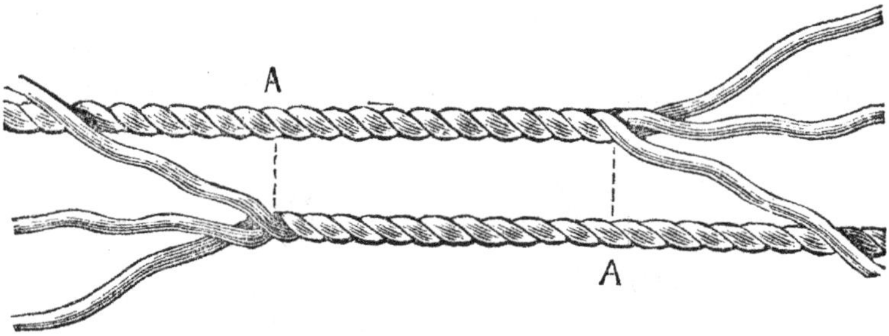

Fig. 118.—Beginning Cut Splice.

marled close down; then clap on a throat and two end seizings of ratline.

In splicing cotton ropes on the Lancashire system proceed as follows:—Short splicing the ends together is the simplest, but would not answer

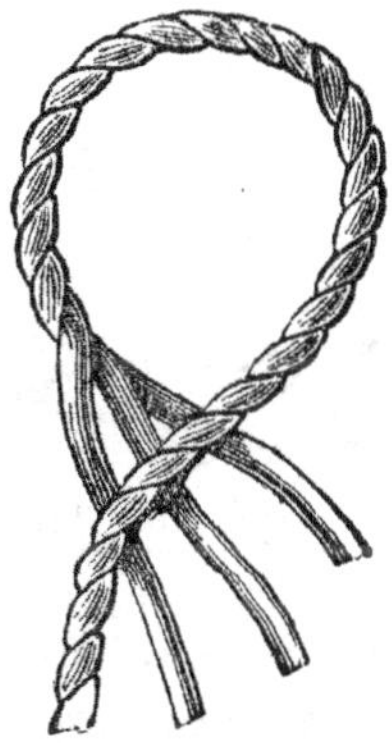

Fig. 119.—Eye Splice.

for running over a small pulley or through the "swallow" of a block; in this case a long splice would suffice. If a small strop only is needed, a one-strand grommet is the neatest. A strand from cotton rope will not keep its lay, or acquired spiral form, and is therefore a very difficult material to work into a grommet. Practising first on hemp is advised. From a piece of rope about a foot longer than the circumference of the intended grommet, unlay a strand with care, to prevent it losing its lay. Now lay up this strand into itself, as shown in Fig. 120. The two ends

Fig. 120.—Splicing Grommet.

will meet each other in the same crevice. Halve these ends, and tie an overhand knot, as shown in Fig. 121, which must not be pulled too tight, or it will buckle-up the grommet (this will be noticed). When this knot is quite snug down in its place, take about one-third of the yarns out of the ends with which the knot was made, and tuck the remaining two-thirds under two strands, missing one (see Fig. 121, where the arrow D shows how to tuck the end A, and arrowhead C shows where end B will come out). Now halve these ends, and, leaving one half, tuck the other as before ; work them in nice and snug, put the grommet

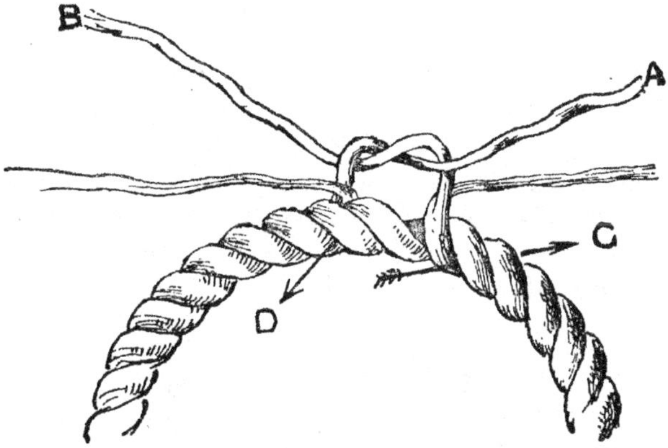

Fig. 121.—Finishing off Grommet.

on a good stretch, and trim off the ends. If the work is well done no joint should be noticeable. Grommets are sometimes finished off by knotting the whole strands, then halving them, and tucking like a common splice. This is not quite as neat, but perhaps a trifle stronger.

CHAPTER VIII.

WORKING CORDAGE.

SEIZING implies the fastening together of two ropes, or different parts of the same rope, with several closely placed coils of small rope, spun-yarn, etc. The several kinds of seizings take their names from the positions they occupy in a ship's rigging. End seizing is a round seizing at the end of a rope. Throat seizing is the first

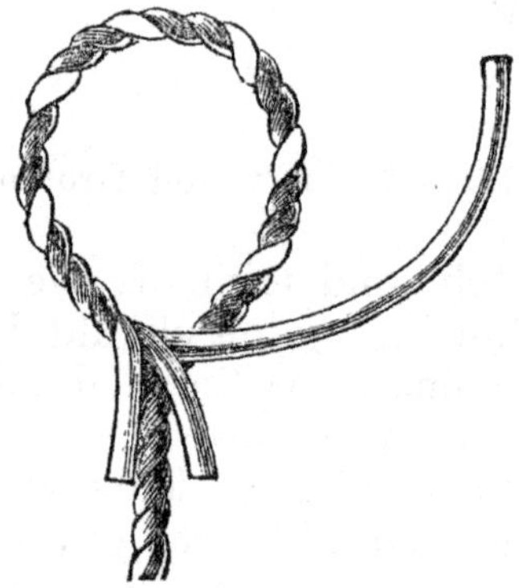

Fig. 122.—Flemish Eye.

seizing clapped on where ropes cross each other. Middle seizing is between a throat and end seizing. Eye seizing is a round seizing next to an eye in the rope.

To make a round seizing, make a small eye in the end of the seizing stuff, and, after taking a turn round both parts of the rope, reeve the ends through the eye, take two or three turns, and haul them taut with a marlinespike hitch (Fig. 40, p. 40); pass eight or ten turns close together, and heave taut. Bring the end back under these turns and out between the last two coils, and pass another series of turns on the top of the others, which are called riders, and are not hove so taut as the first turns. There is always one less of the riding turns than of the lower ones. Two cross turns sometimes are taken, passing between the ropes to be joined and across the whole of the seizing; the end is brought under the last turn, hove tight, and secured, if large, with a wall knot, crossed (Fig. 83, p. 66), and, if small, with an overhand knot, and cut off. Other seizings are done in a similar way.

Sennit is a flat rope, made by plaiting together ropeyarn or spunyarn, the outside yarns being brought over to the middle from each side alter-

Fig. 123.—Grommet.

nately. It always has an odd number of yarns, generally from five to thirteen. French sennit is braided with an even number of yarns passed over and under every other time.

Gaskets are made of braided cordage in the same manner as sennit, and are used for confining the sails when furled to the yards. They are called arm gaskets when used at the ends of the yards. Bunt gaskets are used in the middle of the yard to hold the bunt of the sail, and quarter gaskets between the middle and extremities of the yards.

A Flemish eye (Fig. 122) is a form of eye made without splicing. Unlay one strand at the end of a rope, and bring the two other strands, just as they are, against the standing part, so as to form an eye of the size required. Lay up the unlaid strand in the intervals in the rope from which it has been taken, but in the reverse way—that is, begin at the end and keep on laying it round until it comes down the standing part and lies along with the other strands. The ends are then tapered, marled down, and served over with small stuff.

An artificial eye is the end of a rope unlaid, and the yarns of which the strands are composed are separated. The yarns are now hitched round a piece of wood the size of the proposed eye. They are then marled, parcelled, and served over.

A grommet (Fig. 123) is a ring of rope made by carefully unlaying one strand from a rope and cutting it off. All the turns must be left in it.

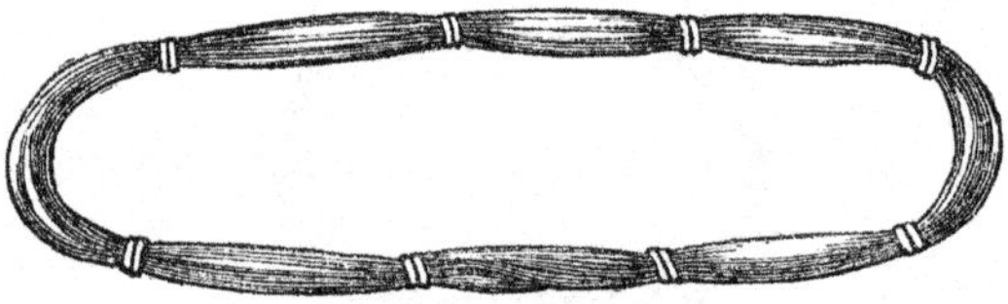

Fig. 124.—Selvagee.

Form a ring by laying one part over the other, taking care that the turns coincide with one another. Pass one end round and round, in the lay, until all the intervals are filled up and the ring is complete. The two ends are secured as in a long splice, first with an overhand knot, and then by dividing the strands and passing half of them under the standing part, and cutting off the remainder. Grommets are used for stropping blocks, handles for chests, snorters for the heel of sprits, etc. They are very often parcelled and served to make them look neater. It is easier to make them if the rope from which the strands are taken is laid up hard.

Selvagee (Fig. 124) is a number of ropeyarns fastened together. To make it, drive two nails into a piece of board at a sufficient distance from one another to form the size of selvagee required. Wind ropeyarn round these to form the desired thickness, and marl them down with spunyarn. They are used to form a neat stropping for blocks, or to go round a spar to which a hook is to be fixed. Fig. 125 shows how a selvagee is employed for fastening a block to a rope. The middle of it is placed against the rope, and the bights passed one over the other until they come close to the rope, when the hook of the block is inserted.

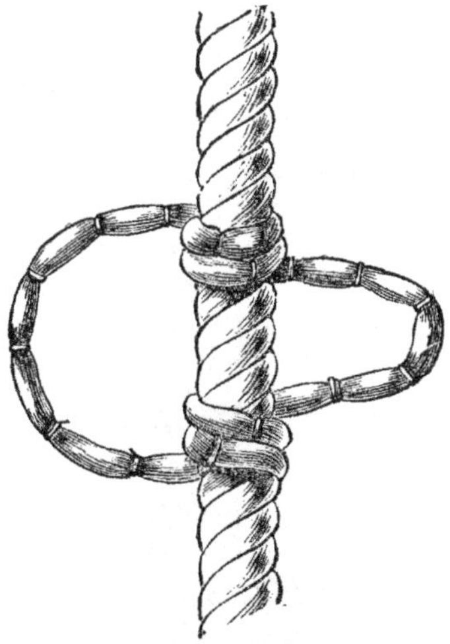

Fig. 125.—Selvagee fastening Block to Rope.

Worming is filling the intervals between the strands of a rope by laying spunyarn or other small stuff into them. This renders the rope more even and smooth for parcelling and serving. The first end of the worming is securely stopped and passed along one of the divisions of the rope. When it has been carried as far as it is required, it is stopped and laid back down another interval, and then forward along the remaining one, and stopped at the end. To estimate the quantity of serving stuff required for a given length of rope, multiply the length of rope to be served by the number of strands in the rope, and add one-third of the product. The result is the length of serving

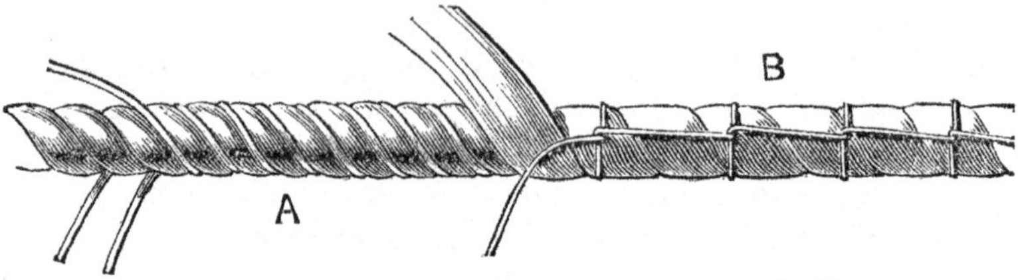

Fig. 126.—Worming, Parcelling, and Marling.

necessary to do the work. Thus, if six fathoms of three-strand rope have to be served:— $(6 \times 3 = 18)$ $(18 \div 3 = 6)$ $18 + 6 = 24$. Thus 24 fathoms is the

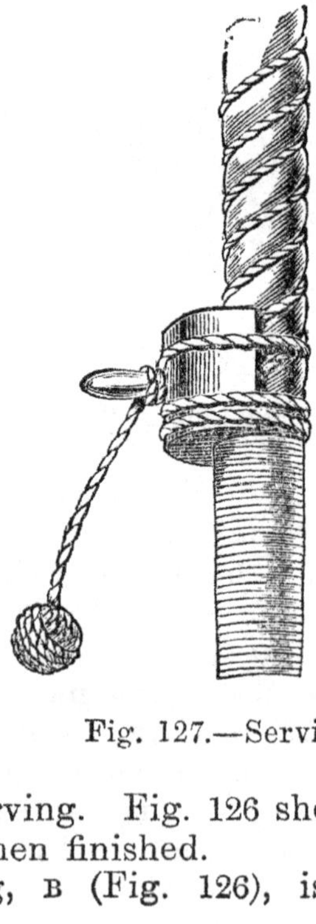

Fig. 127.—Serving.

length of serving. Fig. 126 shows at A how worming looks when finished.

Parcelling, B (Fig. 126), is done by winding

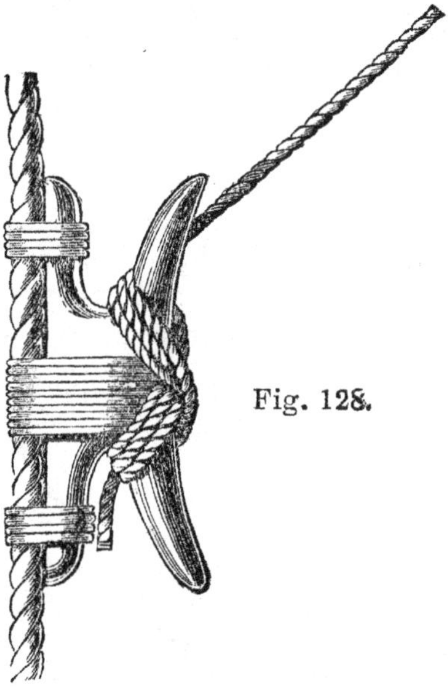

Fig. 128.

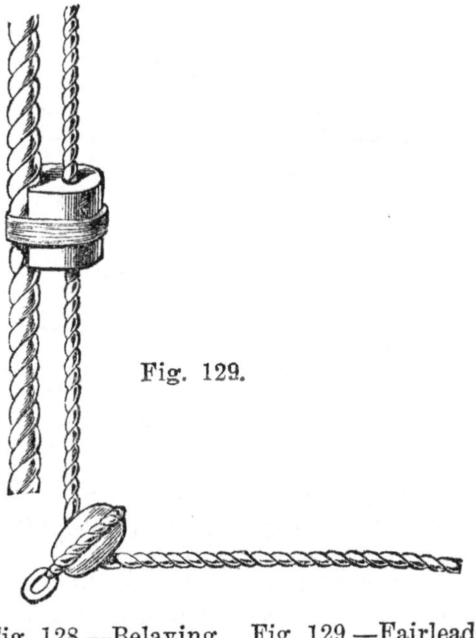

Fig. 129.

Fig. 128.—Belaying. Fig. 129.—Fairleader.

strips of old and generally tarred canvas smoothly round a rope in spiral turns after it has been wormed and before it is served. To secure the canvas it is marled down—that is, some marline or other small stuff is wound round it, which is secured at every turn with a hitch, so that each of the turns is secure and independent of each other (see B, Fig. 126). The proper hitch for securing the turns is shown on a larger scale in Fig. 72, p. 60. In marling down, the coils are

Fig. 130.—Belaying Pin.

not laid close to one another as in serving, but always at some distance apart.

Fig. 127 shows the way a rope is served or covered with coils of spunyarn or other small stuff laid on quite close to one another. The end of the yarn is first secured by placing it under the first two or three coils. A serving mallet (see Fig. 127), after being placed against the rope, has two or three turns passed round the body of it, and another turn or two on the handle. This produces sufficient friction to leave the coils taut as the mallet is worked round the rope by its handle. Another person is required for passing the ball of serving stuff. The service must be put on against the lay of the rope. A rope may be served single-handed by carrying the serving stuff on a large reel, with a hole in its centre large enough for the rope to run through. This is kept just ahead of the mallet, and the serving stuff comes off the reel of its own accord as required. When the mallet is within a few turns of the end, the turns are taken off it by hand, the end is put through them, and heaved well taut.

A rope is belayed or made fast by cross turns round a cleat in the way shown in Fig. 128. The cleat is assumed to be lashed to a stay or other rope, but it is often made fast to some part of the

Fig. 131.—Toggle.

vessel. Occasionally a single hitch is put over the upper horn of the cleat to make the rope still more secure.

A rope is taken at right angles from one part of a vessel to another by means of a "fairleader" (Fig. 129). This is a block of wood with a hole in it big enough to allow the line to run freely through it. The back of it is grooved to fit the rope it is lashed to. Where more than one line has to be led, a piece of board or plank with holes through it is used. A fairleader is not necessarily fastened to a rope, but is fixed in any suitable position.

A belaying pin with a rope made fast to it (Fig. 130) is the usual way of securing running rigging, as it can be made fast and cast off so rapidly.

Fig. 131 shows a method of securing ropes together by means of a toggle, a piece of wood turned to shape and having a groove in the centre, round which the end of a rope is spliced. An eye

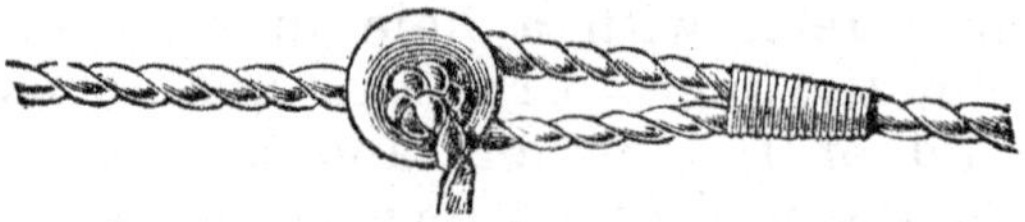

Fig. 132.—Another Form of Toggle.

is made in another rope by any method and the toggle slipped into it. It is undone by slackening the ropes and putting the toggle through the eye, end foremost. Fig. 132 is another form of toggle, the round piece of wood being shaped like a button. It has a hole in the centre, through which a rope is passed and the end knotted.

A fender (Fig. 133) protects the sides of a boat

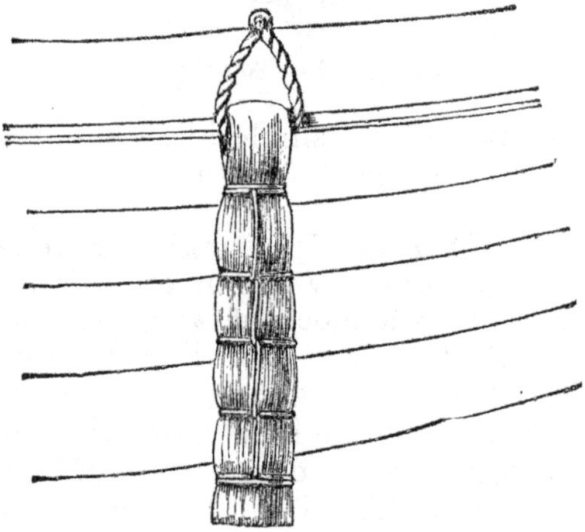

Fig. 133.—Fender on Side of Boat.

from being chafed and the paint or varnish rubbed off. Occasionally a fender is of wood, which is slung over the boat's side by a lanyard reeved through a hole in the end of it, but more often it is of canvas, stuffed with oakum and painted. Fig 133 shows an easily made soft fender which does not need painting. Take a piece of Manilla rope double the length of the fender; unlay it, open the strands, and comb them down until all the yarns lie straight; double it and clap an eye-seizing on it, marling it down as illustrated. A lanyard of small cords, such as log-line, is then spliced into the eye.

Fig. 134 shows a handsomer form of boat's fender. This is made of a centre or "heart" of ropeyarn worked over or grafted with short pieces

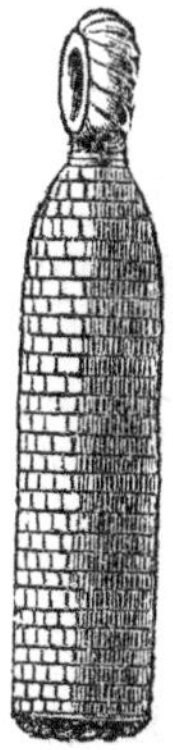

Fig. 134.—Fender with Ropeyarn Heart.

of ropeyarn called "knittles" or "nettles"; this is a kind of weaving. The nettles are first cut to the right length, and the middle part slightly twisted. They are then brought snugly round a thimble, and a seizing put under it. The heart is put into its place and the nettles laid evenly over it. Half of the nettles, taken alternately, are turned back over the eye, the rest lying down the heart. Pass a turn or two of twine or marline, called the warp or filling, round the fender where the nettles separate, and hitch it. The turned back nettles must now be brought down, and those that are down turned up over the eye. The warp is now passed again, and hitched as before. This must be repeated until the whole of the fender is covered with a sort of woven coat as shown in Fig. 134. The ends of the nettles are brought

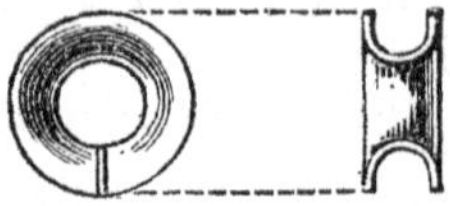

Figs. 135 and 136.—Thimble.

round the last turn of the warp, and interlaced in the grafting. This fender requires a lanyard the same as the last one.

A thimble, shown in side view and section by Figs. 135 and 136, is an iron ring, usually galvanised, and a rope or strop fits snugly in its flanged rim. It may be heart-shaped or round, and is used as a small eye on the end of a rope to receive another rope.

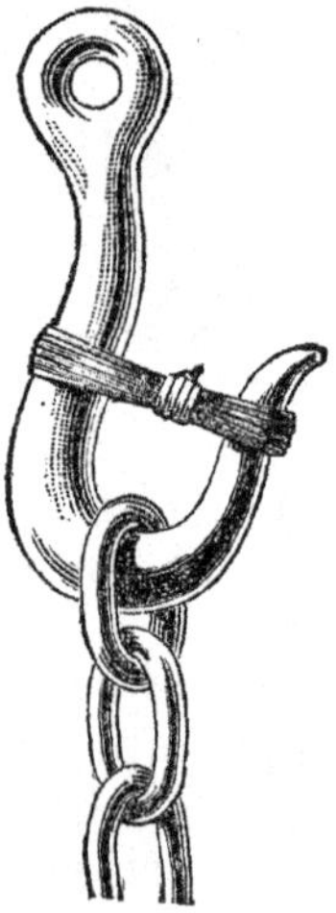

Fig. 137.—Mousing a Hook.

Fig. 137 shows a method of "mousing a hook," that is, preventing a chain slipping off a hook. A few turns of ropeyarn are passed round the end of the hook and the standing part, and the ends brought round the middle a few times, and fastened with a reef-knot.

For "stropping a block" (Fig. 138) a grommet is neat, and a selvagee still neater, especially when leather-covered. The block is first fixed in one bight, so that the lower part of the block sits on the splice, if there is one. A thimble is put into the other bight, and a seizing put on between the block and the thimble, each turn of which is hove taut with a heaver. The turns are, lastly, crossed, and the ends knotted. Fig. 138 shows the work finished. Sometimes the stropping is made

Fig. 138.—Stropping a Block.

by splicing the ends of a suitable length of rope together.

Pointing a rope is done partly to prevent it from untwisting, and partly to make it go more readily through a block or hole. Fig. 139 shows one method. The rope is unlaid for the necessary length, the strands reduced gradually, and then laid up again. The ends are finally whipped with small twine. If necessary the end is stiffened by inserting a piece of stick. Sometimes a "becket"—that is, a piece of small line with an eye at the end—is put into the end and whipped over to secure it, as in Fig. 139. Fig. 140 gives a more elaborate method. The rope is first unlaid, and a stop put on it where the unlaid part begins. As many yarns as are required are taken out and made into nettles by twisting together the two halves of different yarns. The remainder of the yarns is scraped down taper with a knife. Half of the nettles is turned back on the standing part of the rope, and the other half allowed to lie on the

Fig. 139.

Fig. 140.

Figs. 139 and 140.—Pointing a Rope.

scraped part. Two or three turns of twine are hitched round the division of the two sets of strands, and the nettles laid backwards and forwards, the weft being passed each time, as described in making the fender (Fig. 134). The end usually is whipped and a seizing put on the upper part, which is snaked, as illustrated, by passing twine diagonally under and over the outer turns of the seizing alternately—that is, if it comes out over the upper turn, it will go under the bottom one, under the top, and so on until it is finished.

Fig. 141 is part of a mainstay. An eye is first made in the end, and a mouse the shape of a pear raised on the rope with spunyarn. Each turn of the yarn is hove well taut with a large serving mallet, and beaten close. The eye and the rope as far as the mouse are wormed, parcelled, and served over; the mouse and the part below it (the tail) are parcelled with worn canvas, well tarred, and pointed over or grafted with small

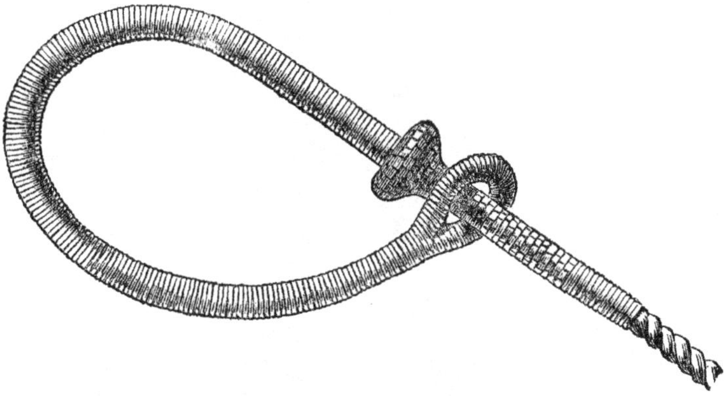

Fig. 141.—Mainstay.

stuff. The mouse is covered with nettles, and their number diminished as they are worked into the smaller parts. Below the pointing it is again served over.

Shear-legs are fastened together as indicated in Fig. 142. They are first laid side by side, and a lashing of rope put round them. The ends of the lashing are carried one up and the other down to form a cross-lashing, and are knotted in the middle. This is called a Portuguese knot. When the legs are separated, the knot becomes very secure.

A neat pair of yoke lines is made and fitted in

the following manner. The length of the lines depends upon the distance of the yoke from the after seat; that distance added to 3 ft. will give a good length for each line. Supposing the top of the back-board to be 3 ft. from the yoke, two-and-a-half fathoms of white cotton rope will be necessary, the extra half fathom being allowed for knotting, etc. Cut this in two equal lengths;

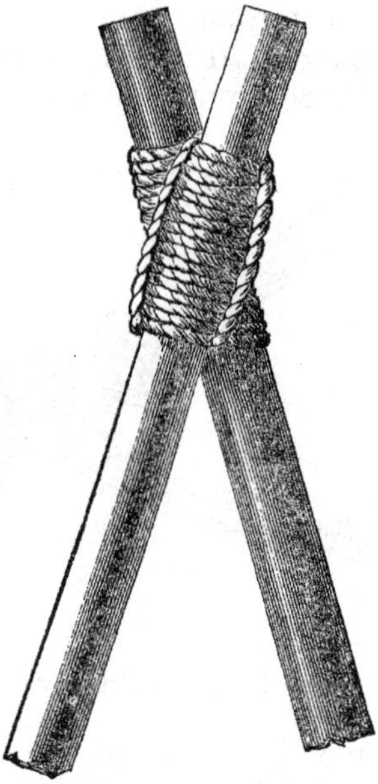

Fig. 142.—Shear-legs.

take one of these and "stop" with a few turns of twine at 14 in. from one end and 6 in. from the other, which will leave 70 in. between the stops or whippings. Now unlay the longer end and whip each strand close to the end, leaving a few inches of spare twine on each, which will be useful when finishing the manrope knot.

Holding the line in the left hand, make a single wall knot, as shown in Fig. 143. The strand A is first placed and held in position by the thumb;

B is next taken and passed round the end of A, then C round the end of B, and up through the loop formed in the first instance with A. Now

Fig. 143.—Wall Knot.

pull together evenly, but not too tightly, and with the three ends remaining work a crown (see Fig. 144), where A is crossed over the knot, then B over A, and C over B, and down through the loop formed by A. After pulling this crown together

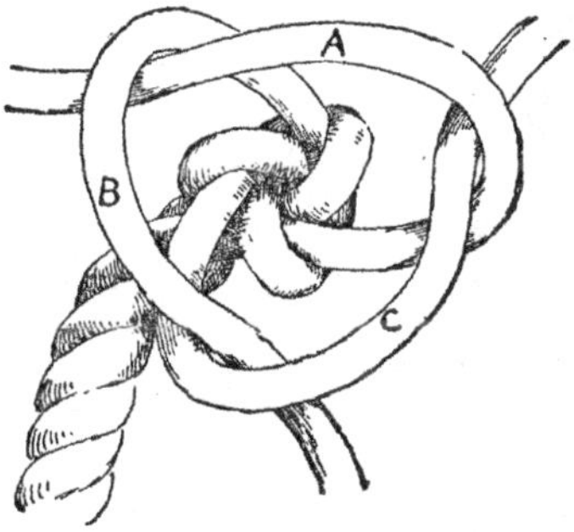

Fig. 144.—Crown.

Fig. 145.—Crown Hauled Taut.

evenly, the state shown by Fig. 145 is reached; notice that the ends lie snug against the strands of the wall knot first formed. Keeping each end on the same side of its partner, let it follow the same round, using a steel pricker to open the way between the strands. The twine ends before-

Fig. 146.—Manrope Knot.

mentioned will now be found useful in coaxing these ends through their holes. When each end has followed its partner round the wall and crown, the ends which will come out at the neck of the knot may be cut off and the manrope knot (Fig.

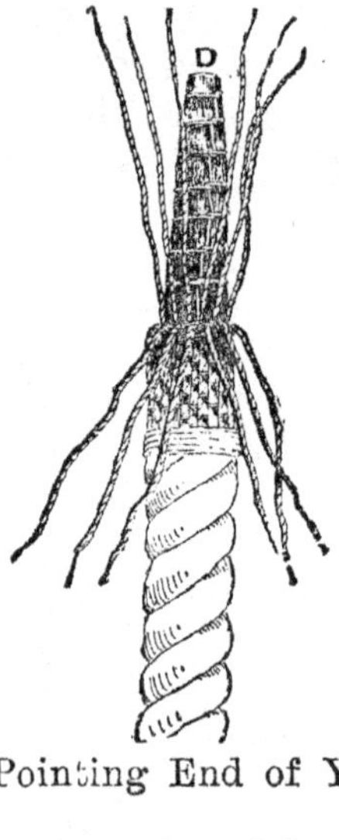

Fig. 147.—Pointing End of Yoke Line.

146) will be completed. A red leather washer with serrated edge is usually pushed up to the knot as a finish.

The other end of the line should be pointed. To do this, unlay and fray out the 6 in. of end, then pick out the yarns nearest the edge—that is, next the whipping—and make a number of 2-ply nettles by laying up these outer yarns, scraping them a little with a knife to make them taper slightly. There must be an even number of nettles, and sufficient to lay close together all round the "heart," which is formed by scraping away the centre yarns to an even taper, then marling down tight with twine as seen at D (Fig. 147). Half the nettles must now be placed along the heart, and

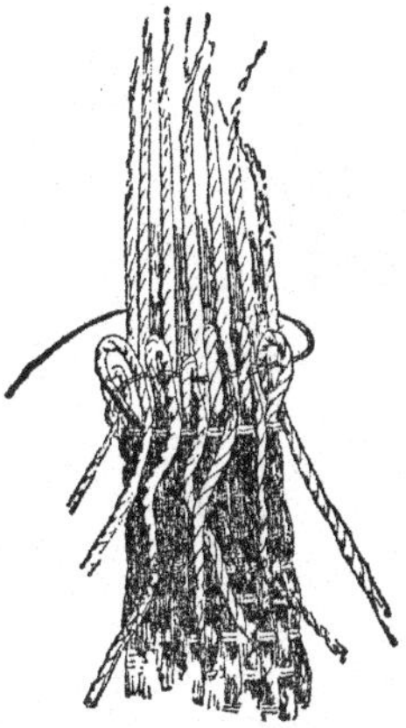

Fig. 148.—Fastening off Yoke Line.

half and half back over the whipping, picking them out alternately. An indiarubber band is convenient to keep them in place while passing the "warp"—that is, two turns of twine hitched round the heart and those nettles lying along it.

The nettles must now change places; all that were placed back must go forward, and all that were along the heart must go back over the whipping, and the warp is again passed twice round and hitched; this process is continued till the point reaches the length of 2 in., or whatever length of point is required. Four times the

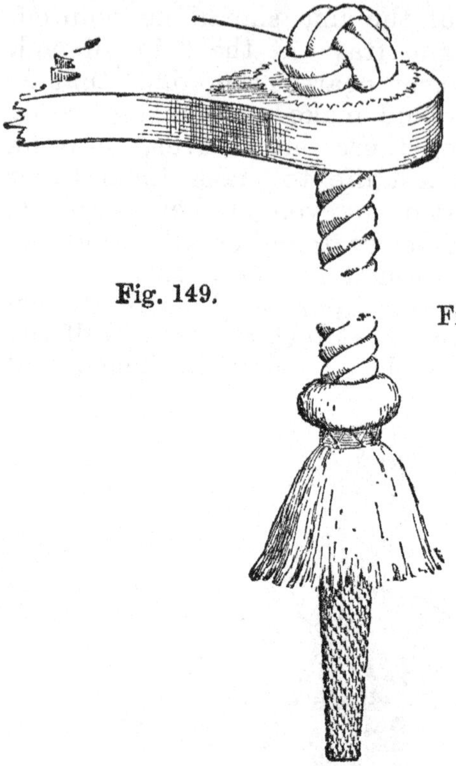

Fig. 149.

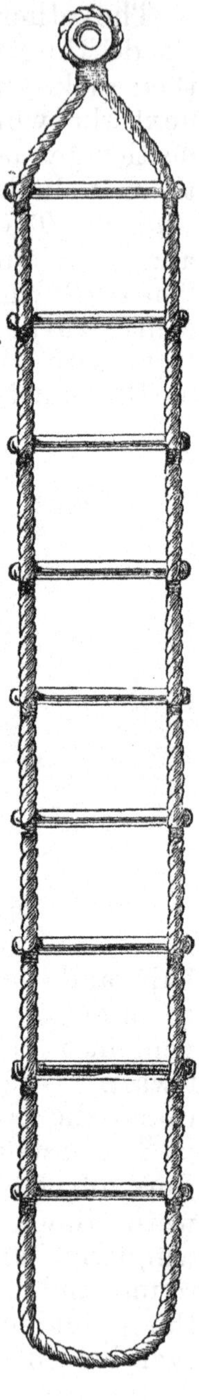

Fig. 150.

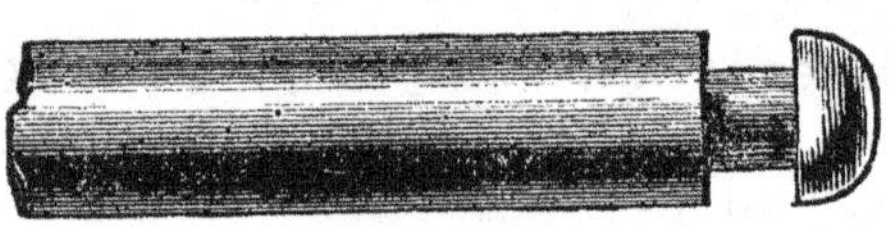

Fig. 151.

Fig. 149.—Yoke Line Complete.
Fig. 150.—Stern Ladder.
Fig. 151.—Round of Stern Ladder.

diameter of the rope is a good rule for this. To fasten off, all the nettles are stroked down towards the tapered end, and all those round which the warp was last passed are doubled back as shown in Fig. 148, three hitched turns of the warp being passed through all the "bites" or loops of the nettles, which are then pulled down and cut off with the other nettle ends. The twine stop is then

Fig. 152.—Single Rope Ladder with Chocks.

taken off and a neat snake whipping put on in its place.

A bunting fringe is generally added as a finish, and this is very easily made by fraying out a piece of blue or red bunting 4 in. square, and laying the threads down together ends even, then placing enough of these round the yoke line to encircle it at about 1 in. from the snake whipping. Pass a number of turns of twine round all near the centre of the bunting threads to form a little knob, then stroke all the bunting threads towards the point and put round all a neat snake-thread whipping just below the knob. This will form a little tassel, as shown in Fig. 149, which represents one yoke ine completed. The other is, of course, made in the same manner.

A rope ladder, owing to its portability, is for some purposes more suitable than one made of wood, especially where the ladder is to be hauled up when not in use. Fig. 150 shows a rope ladder which is made of four-strand rope ; the rounds are turned out of oak to the form shown in Fig. 151. The groove at the ends is for the reception of the

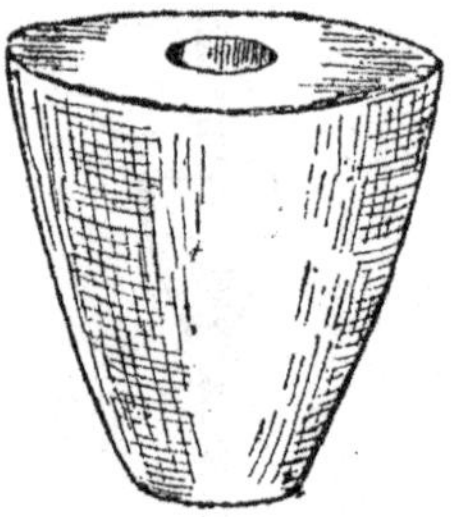

Fig. 153.—Toe Chock for Rope Ladder.

strands of the rope. The rounds are rather more than 1 in. in diameter, and are placed 11 in. apart. The strands are opened with a marlinespike and the rounds inserted between them, two on each side ; a seizing is put on below each round ; a round thimble is put into the upper bight, and an eye seizing is clapped on below it. The lower ends are generally spliced together, or a thimble may be spliced in, as at the upper end, if it is intended to make the lower end fast.

Another form of rope ladder is shown in Fig. 152. A number of oak chocks (Fig. 153) are turned, usually about 5 in. in diameter and 5 in. deep, bored to 1 in. diameter with the grain of the wood ; this will take a 3-in. four-stranded rope. Splice an eye at one end and seize in a thimble for the lashing; put on one of the wooden steps, flat side

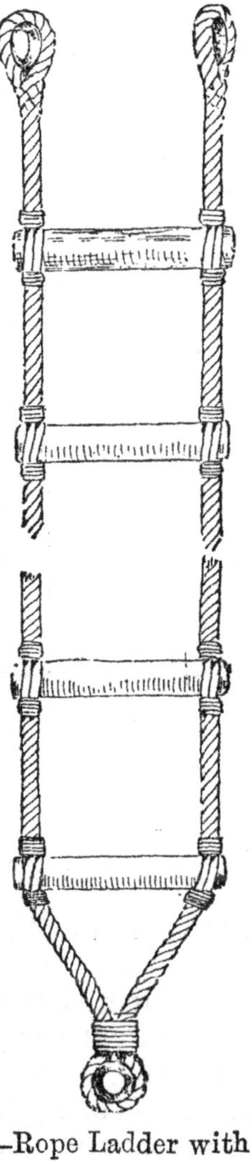

Fig. 154.—Rope Ladder with Rungs.

uppermost towards the thimble and about 30 in. down. Insert a strand of ratline stuff between two strands of the rope, and with the two ends of the strand work a "wall knot" round the rope, then a "crown," which will bring the ends down to the first turns of the wall. Pull the parts up tight, and follow them round once; the result will be a neat knot looking like a two-parted Turk's head. If preferred, two strands may be inserted crossing each other at right angles between the four rope strands; then with all four ends work a double wall knot round the rope. Between the knots, spaced 15 in. apart, slip on the wood. Do

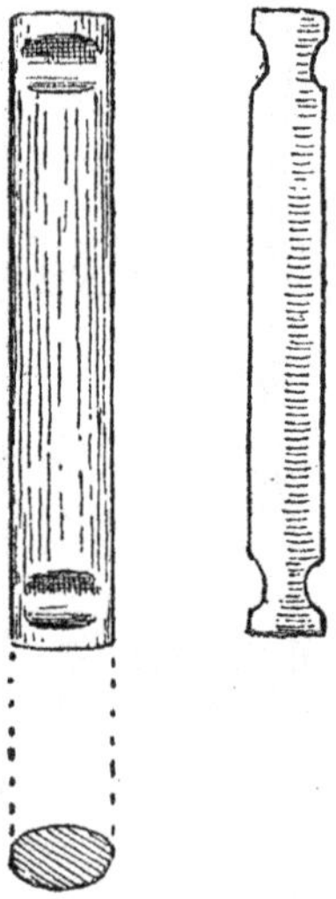

Fig. 155.—Wood Rung.

not cut off the ends short until the knots have got well jammed up.

For a 25-ft. ladder of a better kind, illustrated in Fig. 154, two dozen oval rungs of wood will be required, having at the ends scores rasped as shown at Fig. 155. The rungs may be 12 in. long. Nine fathoms of 4-in. four-stranded rope must be well stretched and the turns taken out, stretched again, etc. Middle it and seize a thimble in the bight, making this fast to a post. Then stretch out both parts together, and mark off the positions for the rungs with chalk, 12 in. apart. At every chalk mark, and round each part of rope separately, put a seven-turn whipping of tarred nettle stuff (this is sold in hanks). Now begin at the end nearest the bight or lower end, open the strands

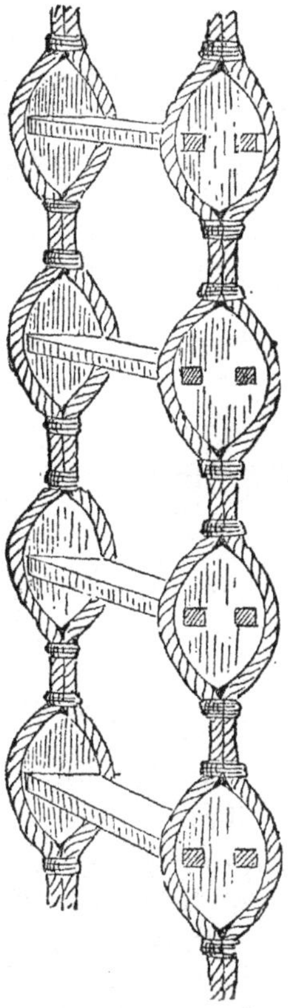

Fig. 156.—Pilot Ladder.

with a fid and spike above the seizings or whippings, and push in a rung, taking care to avoid twisting the rope by inserting the spike between the two strands best situated. When all the rungs are in and the ladder hangs without much twist, the open parts of the strands may be drawn together with a marlinespike and a piece of sennit or by carefully using a screw cramp above the rungs; then a six-turn seizing of nettle stuff put tightly on above each of the rungs keeps them in place. The top rung should be rather stouter than the rest to allow a deeper score. Splice a thimble into each end, and into each thimble a fathom of eighteen-yarn stuff for lashings. This is a good ladder to hang down clear, and is used over ships' sterns and from lower booms, the lower thimble being convenient for boats' "painters."

To hang against a ship's side, where room for the toe projecting over the rung is needed, a pilot

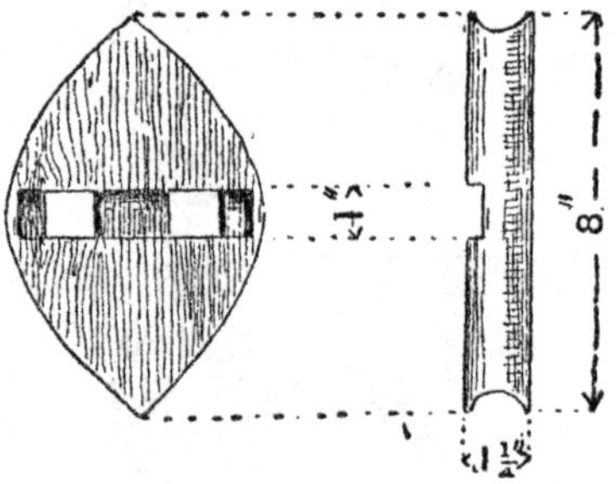

Fig. 157.—Side Piece of Pilot Ladder.

ladder (Fig. 156) is more comfortable, but more difficult to make. The steps are usually of teak, mortised into diamond-shaped side pieces (see Fig. 157) about 8 in. deep by 5 in., with a score rasped out all round the edges. About twenty-three or twenty-four of these steps will be required for a 25-ft. ladder, and eighteen fathoms of 2½-in. three-stranded rope. Prepare by stretching well as before; cut in two equal parts and mark the middle of each length. Place one of the steps immediately over the marks and seize it tight in place by bringing the parts of rope round the score on each side to meet above. If the next step is placed on this seizing and so on, the completed ladder is too stiff to roll up; it is therefore necessary to put two seizings between each step, leaving about $2\frac{1}{2}$ in. of free rope between them. When the steps are thus seized in place, the ends on each side are made even, spliced together, and fitted with thimbles for the lashings.

Mats are used on board ship to prevent chafing. In making wrought mats a piece of small cord is stretched tight horizontally at about the height of a man, and fastened at each end. Across this, hanging by their middles, nettles are placed. These are often made of "foxes"—that is, three or more rope-yarns twisted together by hand, and each rubbed down with tarred canvas or a

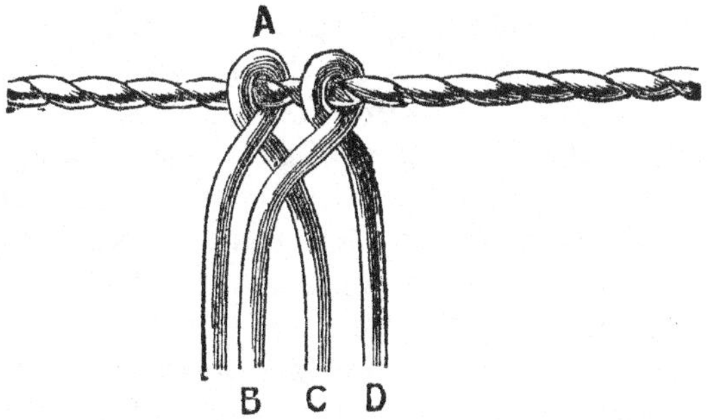

Fig. 158.—Beginning Mat.

handful of rope-yarn. Beginning with the nettle nearest the left hand, it is crossed as A (Fig. 158). Another nettle is then brought up close to the first, and crossed in the same way. The end B is then passed over the strand C, and pushed to the back; another nettle is then brought forward, crossed as before, and one part of it put over the part D, and pressed back. The work is continued this way, working diagonally until a sufficient width is obtained (Fig. 159); then, as no more nettles are added, and the outside nettle on the right is brought over from time to time, a selvage is formed as on the left side. Take care not to twist each of the nettles together at the bottom, so that they may retain their twist until the next in succession is brought down to interweave with them. There is a little difficulty at starting, but afterwards everything goes on easily. Each nettle from the right passes over the next one to it on the left, and is pushed back, the one that has been passed over being taken up first over the next and pushed back as before. Each twist should be pressed tight as it is made. When the mat is deep enough, a selvage is made by straining another piece of cord along the bottom, securing both ends. As each nettle comes down it is half hitched to this, and the next nettle is laid up at the back of it, and so on alternately.

Mats are frequently thrummed; this is done by raising the nettles lying on the top of the mat with a marlinespike or pricker after the mat is finished, and putting short pieces of the nettle stuff underneath. The thrums are then cut off to the same length, and opened out. Of course, wrought mats may be made of any material and used for a variety of purposes. They make most excellent door-mats, and are of everlasting wear. For this purpose they should not be made of tarred stuff. Very pretty mats may be constructed of twine for the foundation and various coloured pieces of worsted for the thrums.

A kind of mat used aboard ship is called a sword mat. It is woven, but a loom is not used. Two small cords, or, better still, two small rods, are secured horizontally and the nettle stuff wound round them, the coils being laid close together. A piece of wood called a "fiddle," as long as the width of the mat and about 2 in. wide and $\frac{3}{8}$ in. thick, has half as many holes bored near the lower edge as these nettles in the mat. Every alternate nettle is secured to this by some twine laced through the holes. Another fiddle is fastened in like manner to the remaining nettles. The work can now be begun. The first fiddle is raised, and the first set of nettles consequently raised with it. What weavers call a "shed" is thus formed—that is, an opening between the two sets of nettles—and along it the weft or filling is passed by means of a netting-needle (see p. 125). The filling is driven well home with a flat piece of wood, tapered towards the edge, called a "sword." The first set of nettles is now allowed to drop, and the second set drawn up with the other fiddle. The filling is passed again and driven home as

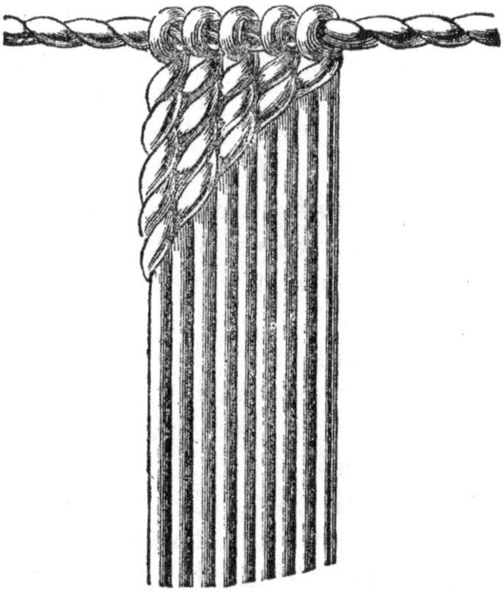

Fig. 159.—Mat Making.

before. The work is thus continued until there is no longer room to use the sword, when the filling must be worked home with a pricker. When the mat is long enough, the filling is fastened off, and the mat is complete. These mats may be thrummed in the same manner as the wrought mats.

A softer kind of mat is made on a foundation of canvas or duck, which is very suitable for the stern-sheets of a boat or any other similar purpose. The material is cut to the right size and folded a short distance from the edge. A hole is made near the selvage with a pricker and a thrum inserted; another hole is then made a short distance from the first and another thrum put in, and so on until the row is completed. Row after row is thus worked until the mat is finished. Of course the holes, and consequently the thrums, go through both parts of the material. When the canvas is pulled straight after each row is finished, the thrums are held securely without any other fastening.

When a pattern is to be worked on the mat, the design must first be drawn on the material in pencil and the canvas folded accordingly; every fold produces two rows of thrums. White duck thrummed with pieces of cotton rope makes very nice, clean-looking mats for boat use, and as they wash well they can always be kept in good order. These mats, with a stout canvas or sacking foundation, thrummed with pieces of untarred hemp rope, serve very well for door-mats, though, of course, they will not last as long as wrought mats.

Very ornamental mats are made somewhat after the same manner as those just described. Any suitable material, of any colour, can be used for the foundation, on which the pattern must be drawn. The mat is folded along the line intended to be worked, and a common pencil laid along the ridge of the fold. The worsted or other material used is threaded in a large needle, and worked over and over the pencil, thus forming, when the pencil is withdrawn, a series of loops on the foundation. Any pattern can thus be worked, provided always that it consists of straight lines. It might be possible to form curved lines by working the loops over the first finger of the left hand, moving the finger after each loop.

CHAPTER IX.

HAMMOCK MAKING.

THIS chapter will describe the netting and slinging of hammocks.

Hammock making requires a netting needle of one of the shapes shown by Figs. 160 and 161. It

Fig. 160.—Netting Needle.

may be made from a piece of $\frac{3}{16}$-in. pearwood, beech, or boxwood about 8 in. long by $\frac{3}{4}$ in. wide. In needles as shown by Fig. 161 the cord is wound round as when filling an ordinary shuttle, and for

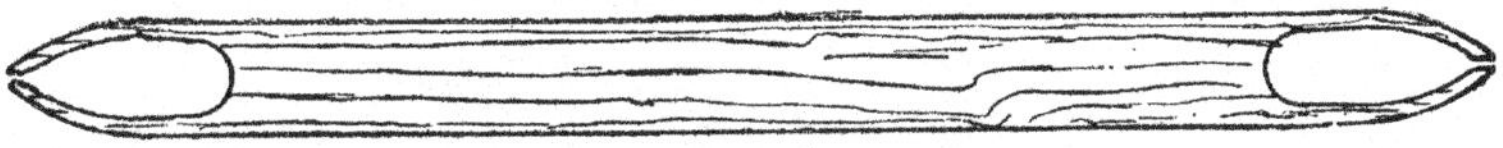

Fig. 161.—Netting Needle.

Fig. 160 the cord is brought round the end at A up one side, round the pin at B, and back the same side, the process being repeated on the other side of the needle

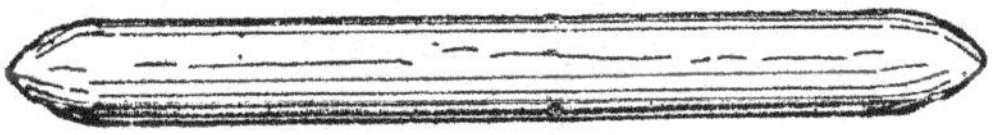

Fig. 162.

Fig. 163.

Figs. 162 and 163.—Mesh Stick.

A mesh stick (Fig. 162) is made of hardwood or bone about 5 in. long and of an oval shape (Fig. 163); it may be about $\frac{3}{4}$ in. by $\frac{1}{4}$ in. in section. At one end of the string to be used for the net tie a loop A (Fig. 164), and place the knot on a nail or hook fixed in some convenient position, as at A (Fig. 165). Place the mesh stick under the loop as at B, put the cord under it, then pass the needle through the loop and pull the cord tight.

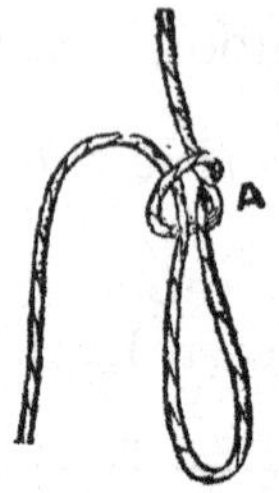

Fig. 164.—Loop in Meshing.

Now place the thumb of the left hand on the cord beyond the loop as at A (Fig. 166), and with a turn of the wrist of the right hand throw the cord to the position shown at B, then pass the needle under the loop C, then through the bight of B, and down as at D, and draw the knot tight, which

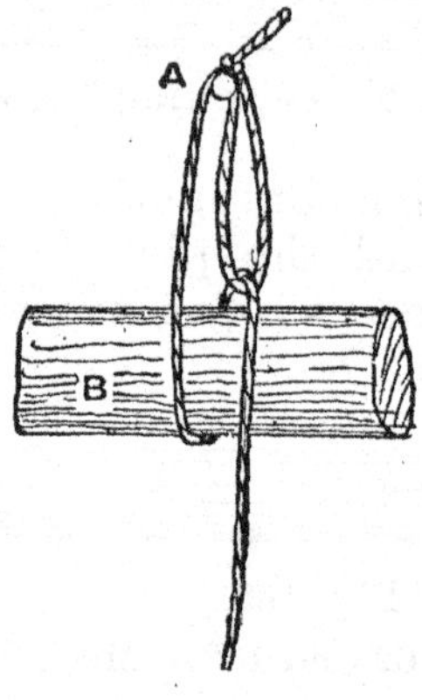

Fig. 165.—First Stage in Meshing.

should then assume the shape shown by Fig. 167. The cord must be held firmly with the thumb at A (Fig. 166) when pulling up the knots, as on this depends the uniformity of the meshes.

To continue the netting the stick is withdrawn and placed under A (Fig. 167), and the needle is then passed under the stick as in Fig. 165, and brought through the loop B (Fig. 167), and the process shown by Fig. 166 is repeated to form another mesh, this being continued to make a chain of meshes, say forty-five or fifty (Fig. 168), sufficient for the width of the hammock. The loop A (Figs. 164, 165, and 168) that was first tied is then untied, and it will then be found that all the meshes are equal in size.

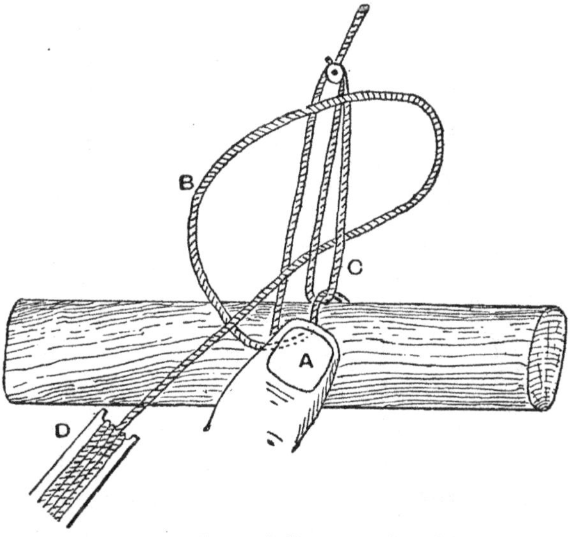

Fig. 166.—Second Stage in Meshing.

Next the chain is opened out at right angles to the line in which it was made, as shown by Fig. 169, and working across is begun by making a mesh at A (Fig. 169), then at B, C, and so on, until the length of the first lot of meshes has been reached, when the net is turned over and another row of meshes worked until the one under A has been reached; then the net is turned again and another row worked, and so on.

The meshes are worked as shown by Fig. 166, but at first, to ensure uniformity, it will be well to put the loops D, E, F, and G (Fig. 169) separately on the hook or nail as the meshes under them are made, but after a little practice a cord may be reeved through the top line of meshes, tied into a loop, and passed over the knee and then over the foot, as the work progresses.

There are three ways of forming the ends.

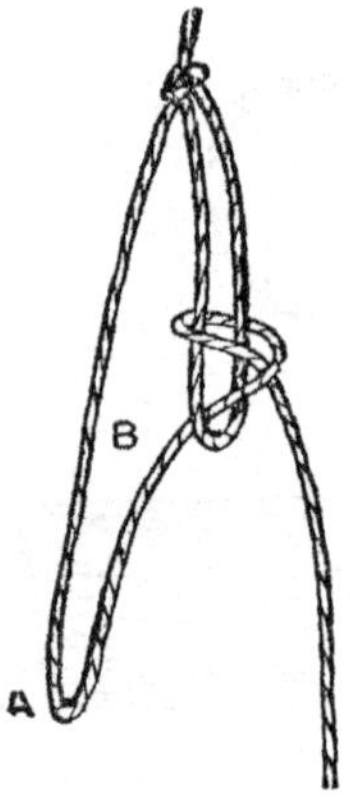

Fig. 167.—Third Stage of Meshing.

Fig. 168.—Chain of Meshes.

An ash stick may be used at each end to which the end meshes are looped and tied, and a piece of codline may be passed through the side meshes on each side and attached to the ends of the sticks. At each end a stout cord is secured to the stick in the form of a triangle for hanging the hammock. The second plan is to tie a number of cords together by doubling them in the centre and forming a loop, and each of the free ends, known as "nettles," is attached to one of the meshes of the net. The third and perhaps the best plan is to reeve a cord about the size of a little finger through the end meshes and splice it into the form of a grommet as shown by Fig. 170. A thimble A is fixed in the end to which the

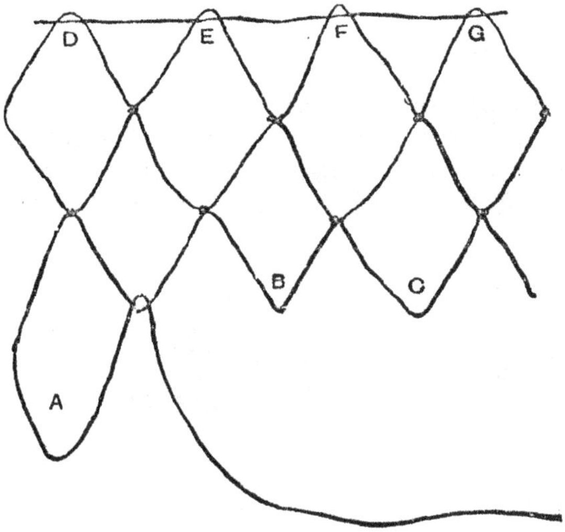

Fig. 169.

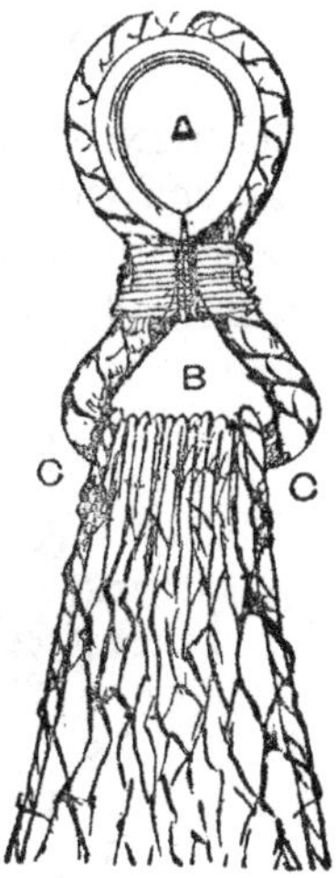

Fig. 170.

Fig. 169.—Beginning of Cross Netting. Fig. 170.—Hammock Clew.

supporting cords are attached, and the cords which are reeved through the side meshes are spliced into the eye B at C. When these "clews" are used the net must be longer than for the sticks or nettles.

CHAPTER X.

LASHINGS AND TIES FOR SCAFFOLDING.

LASHINGS are seizings of rope for heavy work, such as scaffolding; and the lashing of poles, ledgers, and putlogs may be now considered with advantage. Heavy scaffolding is best lashed with chain;

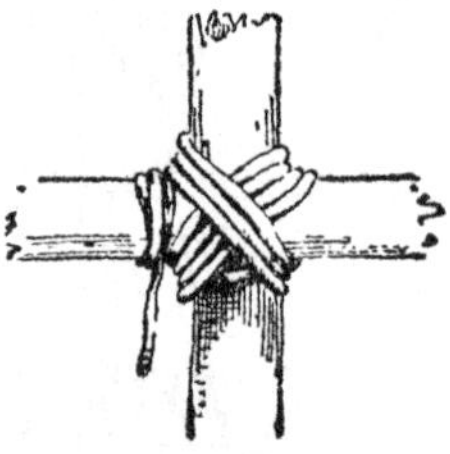

Fig. 171.

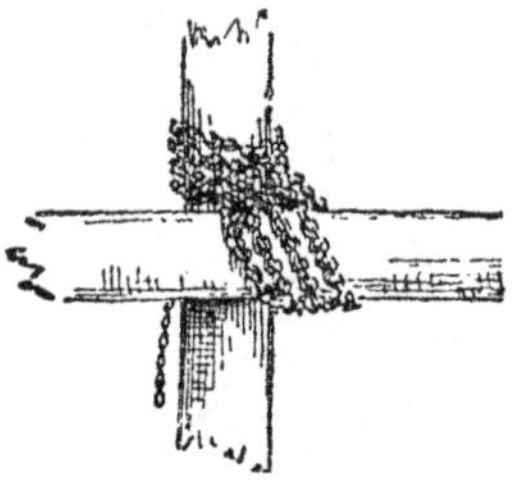

Fig. 172.

Figs. 171 and 172.—Rope and Chain Lashings for Poles and Ledgers.

and it should be noted that, while a rope lashing is hove taut by the steady strain of a heaving bar, a chain requires different treatment, and must be jerked tight with little tugs—an action similar to

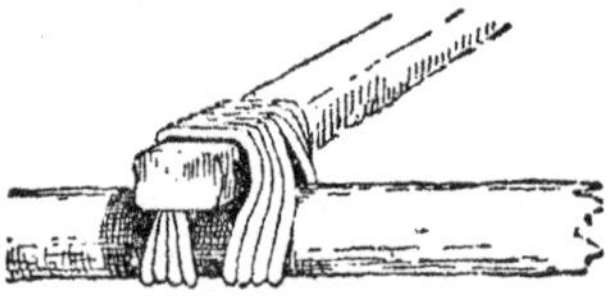

Fig. 173.—Rope Lashing for Putlogs.

that used when shaking out a mat—assisted by an occasional rap with the heaver.

The rope used for lashing all parts of scaffolding should be fairly new, sound, and in size not more than 1½ in. The average length required will be about six yards, and it must be so put on that each turn of the lashing takes its share in the support. To this end the tackle supporting the ledger must not be released until the lashing is completed, otherwise the first turns will be subjected to undue strain.

Fig. 171 shows the lashing of a ledger to a pole; if chain is used a different method is necessary, and it is lashed as in Fig. 172.

Putlogs should be square, or at any rate flat on the upper and lower sides, to prevent any chance of rolling, one end being squared down in size to enter the building in the place of a header brick, the other end being lashed down to the ledger as in Fig. 173. The planks forming the stage are not lashed to the putlogs, and there is great danger in allowing them to project at one or both ends.

Poles are generally about 30 ft. long, and for high buildings it is necessary to lengthen them by lashing on an extra pole. The safest plan is to lash a half pole to the lower part of the first erected; this resting on the ground will form a footing, upon the upper end of which the lengthening pole will stand, being kept in position by three plain lashings (see Fig. 174). Should the lengthening pole be only light, and a footing not be considered necessary, a chain racking should be put on besides two rope lashings.

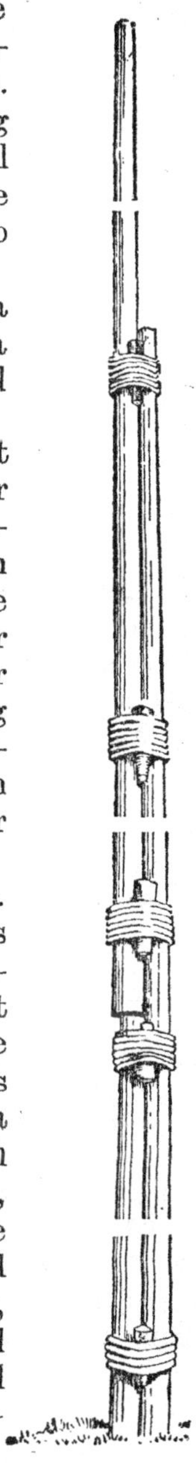

Fig. 174.—Poles Lashed and Wedged, with Footing.

It is usual to wedge scaffold lashings tight, and, provided the wedges are well shaped, it is a convenient way of tightening the work after change of weather. Badly-shaped wedges are apt to cut the rope and sometimes work out.

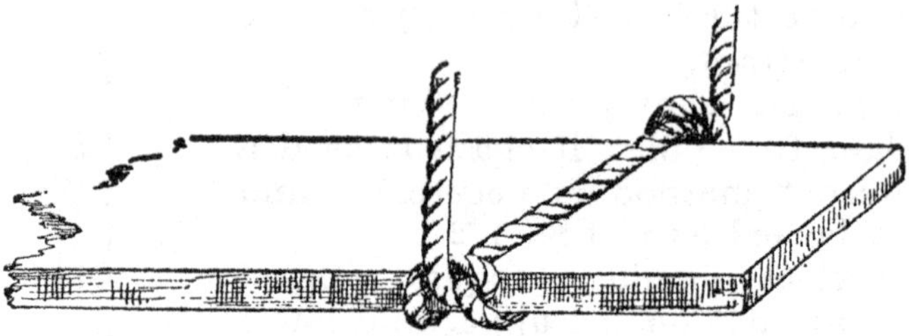

Fig. 175.—Slinging Plank for use as Stage.

Stages for painting, or small repairs on a building, may be quickly rigged up by slinging each end of a plank on the bight of a rope. For this make a marlinespike hitch (see Fig. 40, p. 40), and let the end of the plank take the position

Fig. 176.—Boatswain's Chair.

there occupied by the marlinespike; the double part will be below and the single part above the plank (see Fig. 175). For single-handed work a boatswain's chair, of which an illustration is given by Fig. 176, is convenient.

CHAPTER XI.

SPLICING AND SOCKETING WIRE ROPES.

WIRE ropes are now used to such an extent, and in such a variety of ways, that a knowledge of the best methods of handling them cannot fail to be useful to the many thousands who are brought in daily contact with them. An endeavour will here be made to give such clear and simple instructions as will enable readers, with practice, to execute any job in connection with the splicing of wire ropes. The uses that wire ropes are now put to, and the ways and occupations they are employed in, are so various, that it is almost a necessity for a man to be engaged in their manufacture to have a thorough knowledge of the different methods of handling them. For instance, it would be just as consistent to expect a sailor, used only to eye-splicing, to go to a colliery and put a long splice into a rope, to withstand the enormous stress there applied, as it would be to expect a colliery man to go aboard ship to splice the mainbrace, or any other brace.

Now Fig. 177 is a sketch-plan of an endless band rope, such as may be seen at many collieries in Great Britain. A is the driving sheave on the engine at bank, round which the rope passes three or four times, and leads off to the overhead pulley B. Then it goes down the shaft, perhaps a very considerable distance, under the pulley C, and so on to D, whence it passes round the tightening sheave E, which is movable and contrived with weights to keep the rope always tight, and so prevent surging round the pulleys and sheaves and consequent loss of power. The rope then passes round D′ to the sheave F, the driving of which is the object aimed at, since that sheave in turn may work either one or two hauling ropes, as at G and H, working into the interior of the pit, in addition, perhaps, to a pump. The load on this rope may amount to several tons, and such a rope will therefore require a splice from 60 ft. to 100 ft. long.

Further, no matter what kind of a splice it may be, it must be made well. If it is a ship's hawser, at times the safety of the vessel and all it contains may depend on a splice. Again, an awful disaster might be the result if the splice in a crane rope were to draw.

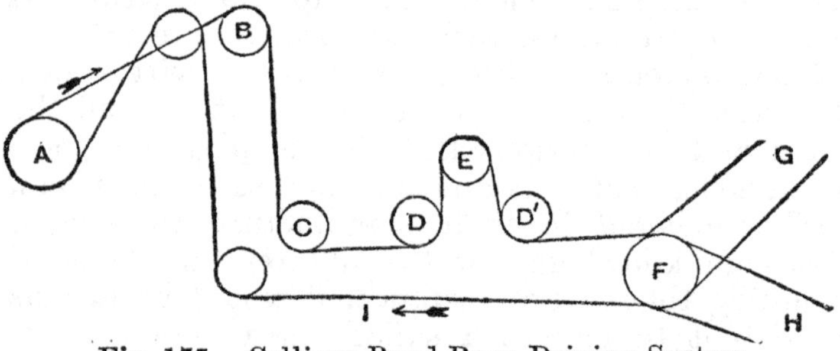

Fig. 177.—Colliery Band Rope Driving System.

There are three kinds of wire rope splices, the short splice, eye splice, and long splice.

A short splice is a bulky splice, and is used only for block straps, slings, etc. It is not suitable for driving ropes or for running tackle, and should never be put into a crane or hoist rope. It is made by unlaying the two ends of the rope to a sufficient length, from 1 ft. to 2 ft., according to size, and interlaying them together as in Fig. 178. Draw them close, and tuck the strands of one under the strands of the other several times. It is only a case of "over and under" as in splicing a hemp rope. After the ends have been tucked a sufficient number of times, about twice, each strand will stand in most cases. Hammer all down snug, and either cut off the ends with a hammer and chisel or twist the wires off one or two at a time, which makes a much neater job.

An eye splice may be made in two different ways—namely, left-handed, or "over and under," and right-handed.

Left-handed splicing is undoubtedly stronger, tuck for tuck, than the right-handed method, for the "bite" on the strands is greater and the frictional adhesion is more acute. The strands and ends are practically plaited, and consequently locked together, and, no matter how the rope unlays itself, the splice is immovable. A splice like

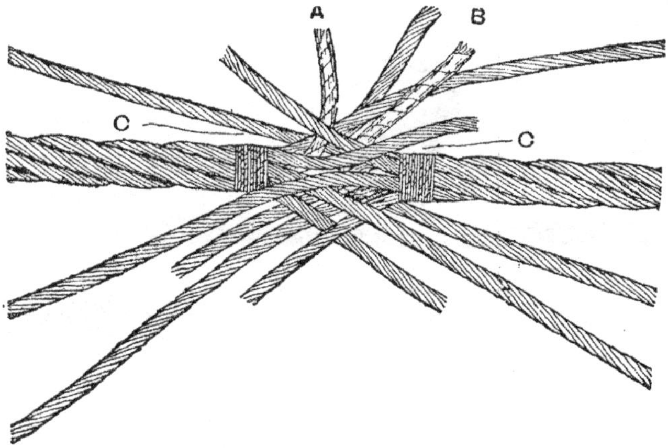

Fig. 178.—"Marrying" a Wire Rope Splice.

this should be put into every crane rope, for frequently the load when lifted from the ground spins round and unlays the rope to a certain extent.

In right-handed splicing, the ends to be tucked are simply laid several times round one strand each, and offer no resistance to any such unlaying process. However, a right-handed splice is easier and neater to make, and these splices are becoming general, some splicers making it a rule to put a "lock" in by tucking the ends once left-handed. A right-handed splice is strong enough for most ordinary purposes, but if a left-handed splice is tucked three times and put on a testing machine, it would break the rope before it would draw out; whereas, if a right-handed splice were tucked three times and put to the same test, it would pull out. However, if the right-handed splice, instead of being tucked three times were tucked eight times, it would stand any stress that could be put on it; in fact, six times through is quite sufficient for ordinary purposes.

The following is the method of right-handed

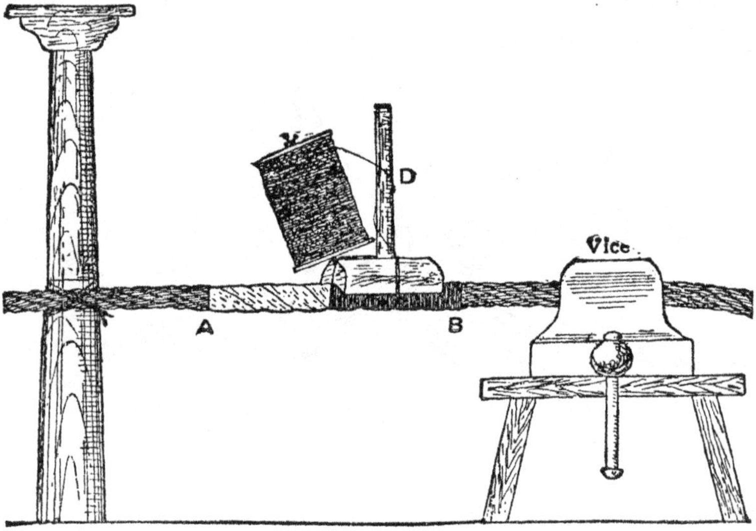

Fig. 179.—Method of Serving Wire Rope.

splicing. First fasten up the rope end from the vice to any convenient column or hook, and mark off from the end, as at A (Fig. 179), the length for the splice, which will be from 1½ ft. to 6 ft., according to the size of rope; say 2 ft. of end for 2 in. cir. rope. Measure with a string round the groove of the thimble, and transfer the length to the rope, marking it as at A B. To put a seizing on the "neck" of the splice, as in Fig. 180, add 6 in. or 8 in. to the length of rope to be served. Now take a narrow strip of parcelling—that is, thin bagging—and bind it neatly round the rope from A to B (Fig. 179). The rule is: Worm and parcel with the lay, but serve the rope the other way. Next take a serving-mallet D (Fig. 179), with a bobbin full of spunyarn upon it, and, beginning at B, serve over the parcelling to A. Cut off and make fast the end of spunyarn, when the work will be ready for turning in the thimble. A chalk mark midway between A and B will serve as a

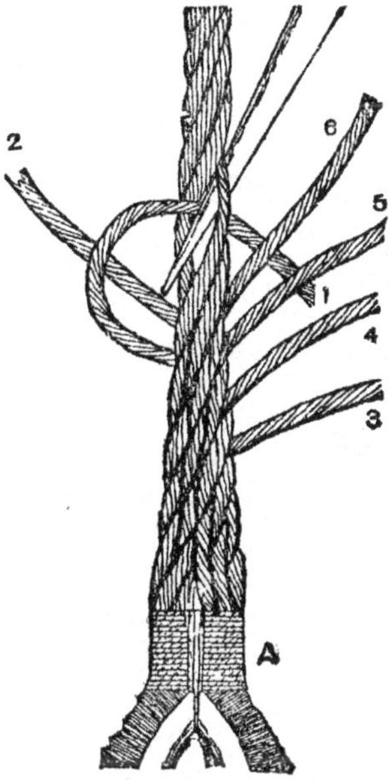

Fig. 180.—Partly-finished Splice.

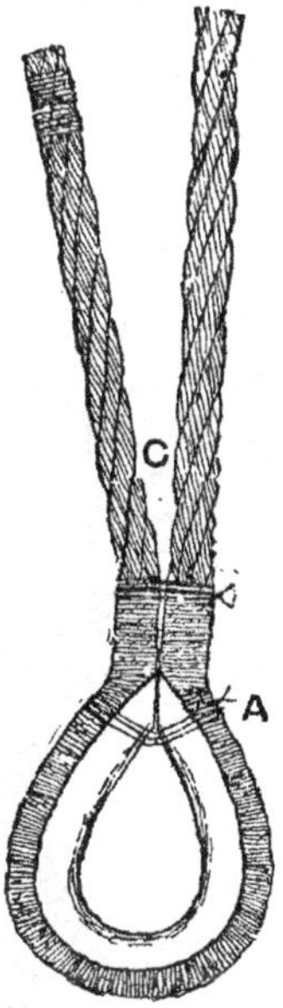

Fig. 181.—Wire Rope ready for Splicing.

guide in getting in the thimble straight. Next bring the end of the rope round on the double to form a loop, with A and B exactly level. Open the vice sufficiently to take in that loop, insert the thimble, taking care to have it exactly straight, and screw up the vice as tightly as possible. Securely fasten the rope and thimble together at the "neck" A (Fig. 181), when the rope should appear as there illustrated. Fasten it in the vice with the thimble hanging downwards, and the part to be spliced held up by a line suspended from some convenient joist or girder. Take the whipping off the rope end, and open out the strands singly, to be ready for tucking. Looking at the rope from the direction of the arrow A (Fig. 182), drive a marlinespike through the two strands on the extreme left, taking care, of course, always to miss the core of the rope. Twist the spike up the

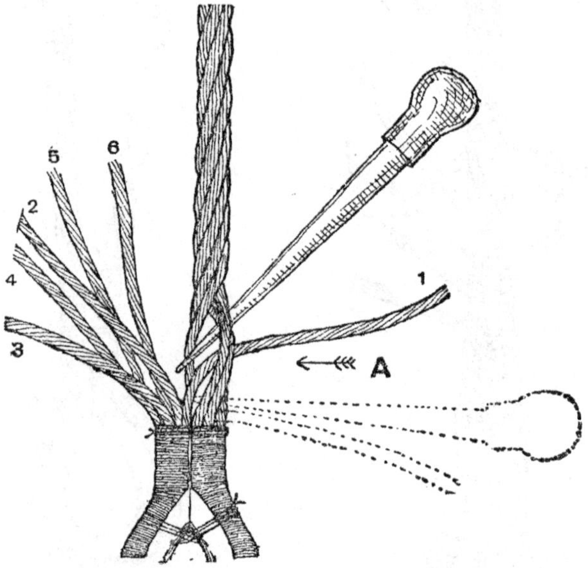

Fig. 182.—Right-handed Splice, First Tuck.

rope about half a turn, so as to make an easier bend for the strand, then take the strand nearest the opening, tuck it through, and haul tight. Then, most important, force down the strand with the marlinespike till it occupies the position indicated in the dotted lines of Fig. 182.

At the next tuck, with the spike take in the strand to the right, along with the two just gone through, making an opening through three strands, with the point of the spike coming out in the same place as in the first tuck. Insert the next nearest end, strand two, and work in snug as before. Of the three strands the spike has just been through, take the one nearest to the left, and drive the spike in so as to make the point come out at the same opening as before ; tuck in No. 3 strand, and that will make, as in Fig. 183, three strand ends entering into the same opening in the rope, but coming out between different strands. After this,

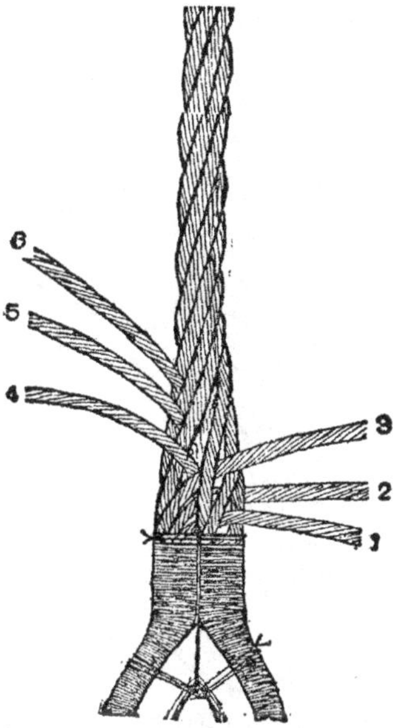

Fig. 183.—Right-handed Splice, all Ends Tucked Once.

simply keep on taking the next strand and the next end, following round to the left, till all the ends are tucked once, when an end will come out between each pair of strands.

The rest is easy. Merely keep on repeating the operation, with strand and end, until the splice is long enough and strong enough. As Fig. 180 shows, after the first set of tucks, it is only a case of each end twisting and re-twisting round its own particular strand, all the way up. It merely thickens the strands. After the ends are tucked three or four times, it will make a neater splice if the ends are split and the splice is tapered at the finish by leaving behind one-half of each end, while the other half is tucked once or twice more. Having finished the tucking, take the splice down from the vice, cut off all strand ends quite close, and hammer all down snugly. At the neck put on the

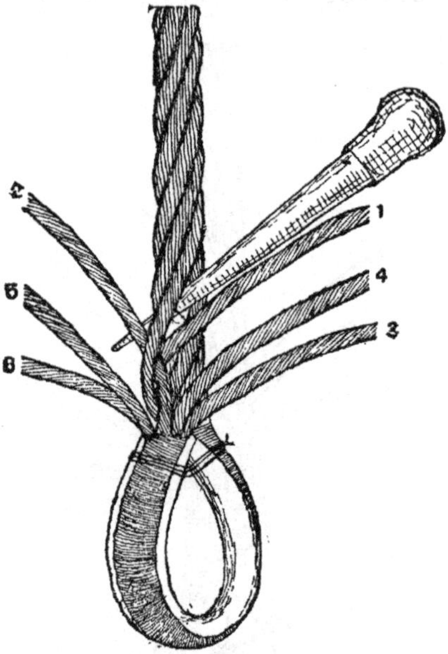

Fig. 184.—Left-handed Splice, First Tuck.

seizing wire, if any, parcel and serve, and the right-handed eye-splice is complete.

Seizing wire is really a seven-wire strand, made of soft wire, about No. 18 or No. 19 gauge. It is put on a splice for the double purpose of strengthening the splice and rendering it easier to take a weak or broken thimble out of the eye, to be replaced by a stronger one. It is simply bound tightly round the two ropes at the neck, and the end is brought up and round the middle twice or thrice and made secure, as at A (Fig. 180).

Left-handed eye-splicing should be comparatively easy if the preceding instructions on right-handed eye-splicing have been carefully followed. Although it is not essential, it makes a neater and a closer splice if a slightly different method of starting is adopted. In splicing a thimble into a crane rope or a trawl-warp, two ropes in which it is very advisable to put left-handed splices, it is

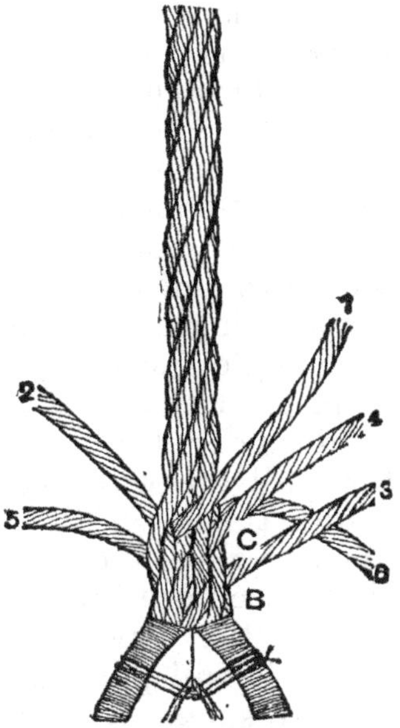

Fig. 185.—Left-handed Splice, all Ends Tucked Once.

unusual to put any seizing on the neck; instead, the rope is spliced right down to the thimble. Serve sufficient rope to go round the thimble only, leaving out sufficient end for the splice. Turn in and screw up very tight in the vice, make fast at the neck, as before described, fasten the rope up with the thimble hanging downwards, and open out the ends.

Different splicers have different ways of starting, but a very good method is as follows: Turn to Fig. 181, and, taking the strand in the rope nearest the point marked C, open it out with the spike. Select the end that is nearest to it, which will be the inside middle one of the bunch, bring it through to the right hand, as in Fig. 184, and force it down well. Still keeping the spike through the same strand, select the middle strand end, on the outside of the bunch, and put it through the same aperture, but in the opposite direction—that is, left-handed. Of the four strands now remaining, first take the two at the right, and put each end through its nearest strands, right-handed, as at B C (Fig. 185), and the other two at the left, and tuck them through their nearest neighbours, left-handed. There is now an end coming out between each strand, and if the rope is now taken down and hammered well at this part, a very neat starting is formed All that remains now is very simple; just keep working round to the left, tucking each end as it is reached over its neighbour and under the next one. Work "over and under" all the way, until each end is tucked three or four times.

The appearance of the splice will be improved if the ends are halved before taking the final tuck.

If the strands of the rope to be spliced are very hard and stiff, it will perhaps be difficult to pull in the ends quite snug. To get over this difficulty, before tucking give each strand a sharp bend close to the rope, in the direction in which it has to be tucked, when it will spring into its place with comparatively little pulling. When all the ends are tucked, hammer the splice well, as that makes the wires knit more into one another, and consequently increases the "bite." All that remains now is to cut the ends off, and serve over, finishing at the thimble. It is a good plan to take the splice down now and then during the process of tucking, and hammer it well, as that serves to drive out all the slack at the ends.

Long-splicing is undoubtedly the most important form of splicing. By its use two pieces of rope may be joined together, or a rope may be made endless, without increasing its thickness at the splice. In fact, none but a practised eye can discern where a well-made splice is, after it has been made and set running, as all ends are completely hidden. Ability to splice well in this style commands many good jobs at collieries in this country and in South African mines.

Of course, in long-splicing, as in many other things, different men have different styles. Suppose, for example, a colliery hauling rope is to be spliced, and that it works endless from a hauling engine to a terminal return pulley, mounted on the tension bogie; suppose also that the rope has been put in position, with the ends left at the most convenient place ready for splicing. If 60 ft. of end is available for the splice, decide on that length, and, measuring about 40 ft. from each end, make the rope fast to the rails at one side, and at the other side fix a block and tackle, and haul in every bit of slack lying on the hauling road, taking especial care that the tension bogie is pulled right up to the top of the tension "ways." This is a very important consideration, because if, through the rope stretching, the tension pulley got down to its limit before the rope was worn out, it would necessitate cutting the rope and making a new splice, or opening out the old splice and re-splicing there again after shortening the rope.

Where practicable, have the tension of a sufficient length that by the time it gets to its limit there is sufficient length of rope to make a new splice, in case the old one is found to be giving way. Some ropes stretch more than others in working, and it is difficult to say accurately how much per cent. a rope will stretch, as much depends on the way the rope has been made. However, the larger the hempen core inside the rope or the shorter the lay or spiral twist the more the stretch. It is wise, therefore, when splicing a new rope never to give any slack away.

Having got the rope hauled as tight as possible, carefully measure 30 ft. from each end of the rope, and there tie a strong whipping. It is important that these whippings should be put on each rope at a distance which shall ensure that they will easily come quite up to each other when the two ends are joined together. Next, take the whippings off the extreme ends of rope and open out the strands. Some splicers do this in pairs; others simply halve the rope, opening it out in two bunches of three strands each. A better way, considering laying them in again, is to open out one strand, then miss one and open out the next to that, then miss another and take the next one again.

There are now three single strands opened out to the whipping at 30 ft. from the end, and between each of these strands there is one strand unopened. This work must be done at each end of the rope, of course. Taking these two bunches of three unopened strands, cut them off about 6 in. from the whipping and throw them on one side. Open out the 6 in., obtaining three strands 30 ft. long, and between each pair is a strand 6 in. long, with the same length of core projecting through the centre of the rope. Join the two ends together as in Fig. 178 (p. 135), taking care in every case to have the strands placed so that a long strand will fall in where a short one comes out, and at the same time pull out between the strands the two short pieces of core A B. Place helpers on each side to pull on the long strands, as in a tug of war, and after cutting the two whippings C get the helpers to pull in opposite directions, until the two ropes are jammed close up to each other, with the strands ready to fall naturally into the lay of the rope. If this is not done properly, or if the rope is allowed to slip back, the appearance of the finished splice will be spoiled by an unsightly long place in the lay.

Next clamp or securely tie the three long strands at one side to the rope, and proceed to lay the other three long strands into place. Select a long strand and the short one that touches it, open out the short strand, and lay the long one in its place. Keep on doing that until all but 5 ft. of the long strand has been laid in; then stop and lock these two strands together by crossing them, so that they will not open out. Now measure back 5 ft. on the strand just opened out and cut the rest off, as only 5 ft. will be needed on each end. Open out another short strand, at the same time laying in a long one, to the length of 15 ft. only, and the next long strand must be laid in just 5 ft. Now release the other three strands made fast to the rope, and treat them in the same way, laying in a long strand where a short one is taken out. Make all the ends 5 ft. long, putting a small whipping on each before cutting to prevent opening out, when the splice will be as in Fig. 186.

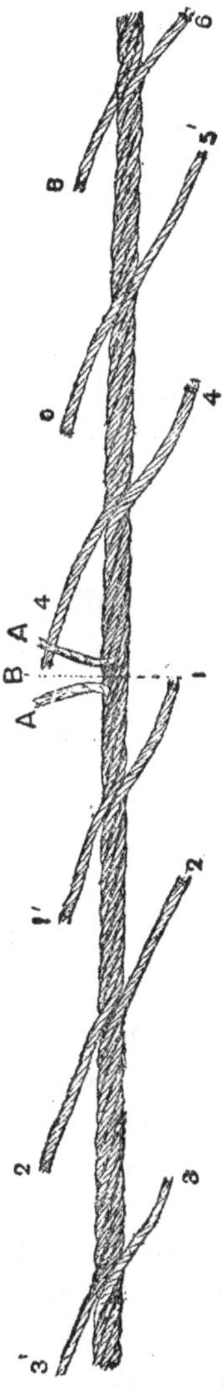

Fig. 186.—Long Splice, all Ends ready for Tucking.

There are now twelve strand ends, 5 ft. long, which must be worked into the inside of the rope and completely hidden; but before this some lengths of thin bagging must be prepared, about 2 in. wide, or, failing that, get some thin yarn, which must be wrapped securely and evenly round each strand for about 2 ft. or 3 ft. from the end. Strips of bagging are the better, and can be expeditiously put on by getting each helper to wrap an end. This gives the outer strands something to knit or bed into when the hempen core is taken out of the centre of the rope and the steel strand ends are worked in. After considering the difference between the size of the hempen core and of the outer strands, necessitated, of course, by the

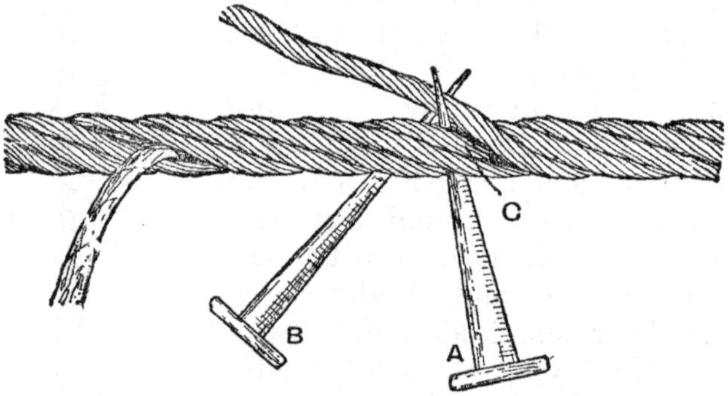

Fig. 187.—Making Round Joint.

spiral form of the outer strands, it will be obvious that the "bite" of the outer on the inner strands would be very little indeed if the ends were worked in bare. When all the ends are wrapped and cut to such a length that they just butt against each other when worked inside the rope, the work is ready for tucking. Take hold of the piece of hempen core A (Fig. 186) and pull it out just to past the first joint—that is, the two ends 1 and 1′. In all probability, if the rope is large or rather hard laid, this will be difficult, unless a tapered round spike is driven through the two strands immediately behind it.

Strand 1 has now to be worked inside the rope where the core came out. To do that, take a flat tucking spike and drive it through the strands 1′ and the next one with the point coming out and covering strand 1. Twist the spike round in the lay of the rope, and, of course, towards the original joint B in the splice; but so manipulate the spike, or "tucker," that strand 1 will fall into the inside of the rope immediately the worker starts to twist, as at C (Fig. 187).

When working a strand from the outside to the inside of a rope it should be done short, sharp, and at once. Good joints must be made short, for then the twists in the strands so unite with one another as to make it appear as though they were blended into one strand.

The "partner" strand to the one just worked in

Fig. 188.—Half-round Top Swage.

—that is, strand 1′—must next be tucked out of sight, but before that can be done it will be necessary to consider that if the wires in the strand are spun or twisted in the opposite direction to that in which the rope is "laid" or "closed," then an ordinary rope is formed. But if the wires and the strands are both twisted in the same direction there is formed a make of rope commonly called a "Langs" or "Albert" laid rope.

In Figs. 179 to 185 ordinary ropes are illustrated, but in Figs. 178, 186, and 187 Langs ropes are shown.

In an ordinary rope the strands are tucked into the inside, side by side at the joint, which is known as a "flat joint," while in the case of a

Langs rope the strands are crossed over each other at the joint, and this is known as a round joint.

Fig. 187 also illustrates a good way to place the spikes to work in strand 1′ and make a round

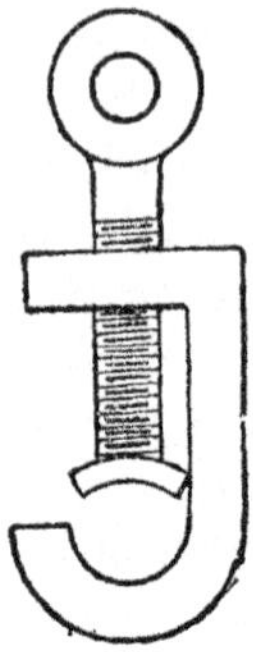

Fig. 189.—Wire Rope Clamp.

joint. Each spike is driven through two strands, and, as shown, A is the leading spike which tucks the strand into its place, while B is used simply to combine with A in making the joint and forcing the strand 1′ into the centre of the rope. This is effected by placing the strand between the points of the two spikes and twisting them up in opposite directions. If the spike A is twisted up towards the worker, and spike B in the opposite direction, the obvious result is that the points of the two spikes come down, and simply crush strand 1′ into the centre of the rope. Then go on twisting spike

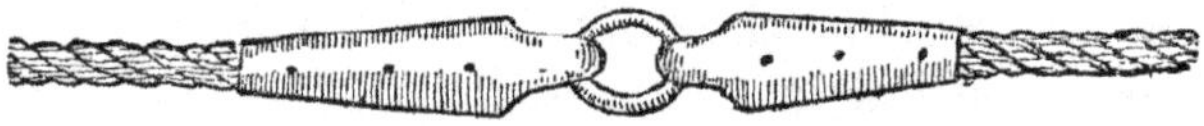

Fig. 190.—Link Socket.

A along the lay, and spike B can be pulled out as soon as the worker gets away from the joint.

If these joints are closely and neatly made, each in its own way, the two strands forming the joint knit into each other's lay or twist, and will be scarcely noticeable. The above operation must, of course, be repeated until all the ends are tucked into the inside of the rope. Tuck them in as numbered in Fig. 186, namely 1, 1′, 2, 2′, and so on. Be very particular in seeing that the

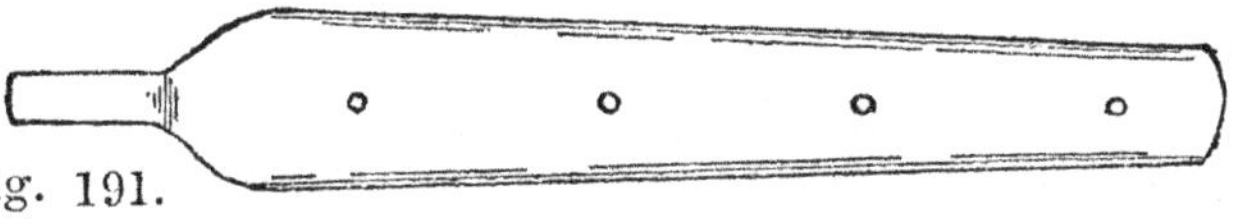

Fig. 191.

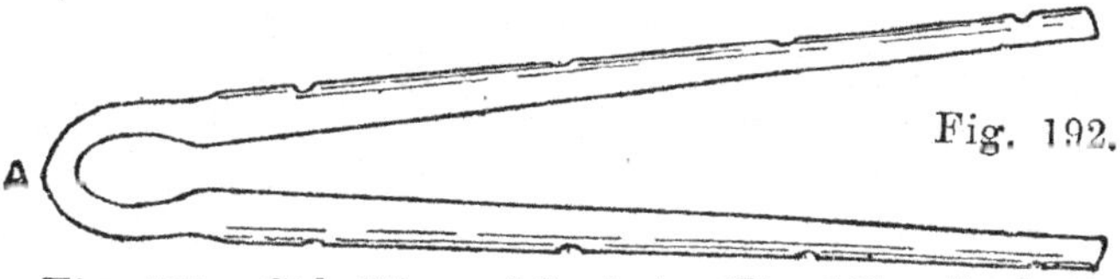

Fig. 192.

Fig. 191.—Side View of Socket. Fig. 192.—Socket Opened Out.

ends exactly butt each other, or a lump results on the splice; if they do not butt, the outer strands will sink in. When all the ends are properly tucked, the long splice is practically completed, but it will improve matters to round off all joints and uneven places with a half-round top-swage, about the same diameter as the rope.

About all the tools necessary for long splicing are two flat spikes or tuckers and one round one, each say 18 in. long, a sledge and set for cutting the rope and strands, a hand hammer for driving in the spikes, a strong pair of cutting pliers and

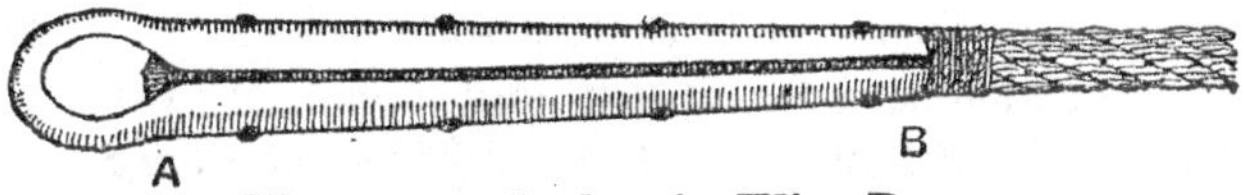

Fig. 193.—Socket in Wire Rope.

the top-swage (Fig. 188). A very useful wire rope clamp is shown in Fig. 189.

In addition to the method of fitting attachments to wire ropes by means of splicing, as already described in this chapter, there is the method of socketing, shoeing, or capping, as it is variously called. There is a great difference between the rough product of the colliery blacksmith, in the shape of the common link socket, and the highly finished cast-steel socket or the turned steel socket of the engineer; and in concluding this chapter it is purposed to deal with them all.

Fig. 190 is a view of the ordinary link socket, often used at collieries as a means of temporarily repairing a broken rope. A whipping of soft wire or spun-yarn is put on each end of the rope, and a few wires then bent sharply back over the whipping and cut off an inch or two shorter than the socket that fits over them. The rest of the rope-end is cut off as close to the whipping as possible,

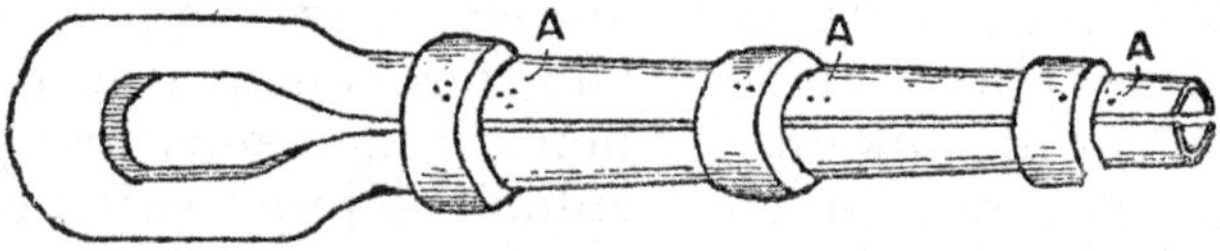

Fig. 194.—Hooped Winding Rope Socket.

and the sockets, which, of course, are ready opened, are put on, hammered down close, and firmly clinched with three rivets.

A very common socket in everyday use is shown by Figs. 191 to 193; this is principally used on haulage ropes and ships' steering-gear. The eye may be of either round or square section, the latter being the strongest and best. As a rule, these sockets are fixed on the rope by means of rivets only, but when attached to winding ropes they are secured with strong steel hoops, which are forged to make a close fit at intermediate distances on the socket, as in Fig. 194.

Sometimes hoops are used in conjunction with rivets. When a hooped socket leaves the blacksmith's shop, the hoops should be a close fit at equal distances on the socket; and as a guide to replacing them correctly, both socket and hoops should be pop-marked with a centre-punch as shown.

The method of preparing the wire rope for these sockets is the same whether hoops or rivets are used. First put a strong wire whipping on the

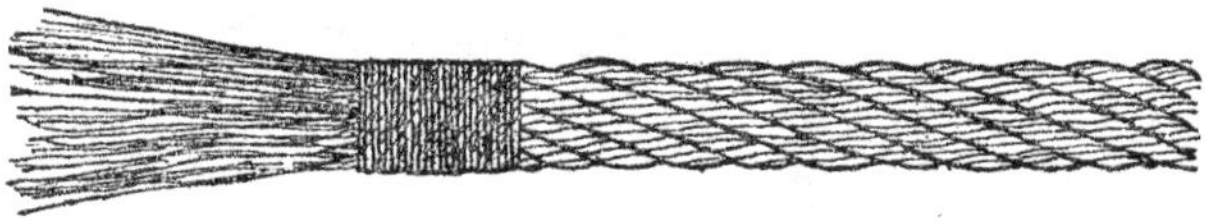

Fig. 195.—Preparing Bulb on Wire Rope End.

rope at about the length of the socket from the rope-end. Untwist the wires of the part left over and straighten them out (see Fig. 195); then bend them sharply back, one or two at a time, over the whipping and over the rope. If the rope has a wire core in the centre, the wires forming it should be bent back the same as the outside wires; but if the core is of hemp, it must be cut off close to the whipping. After hammering snug and close, all these wires have to be cut to a different length, so as to form a tapered bulb corresponding to the tapered cavity inside the socket. Cut the longest wires an inch or two shorter than the length of the socket, measuring from A to B (Fig. 193), and gradually make the other wires shorter and shorter all the way towards the end. These

Fig. 196.—Finished Bulb on Wire Rope End.

wires must now be tightly and closely served with soft copper wire or spun-yarn, using the serving mallet already described in this chapter. For common work spun-yarn is used, and for the best class of work soft copper wire.

In bending back the wires, care must be taken

to ensure their being bent equally all round the rope and not merely at the top and bottom or on one side only. The object is to form on the end of the rope a bulb which will completely fill the cavity inside the socket. To do this, the operator

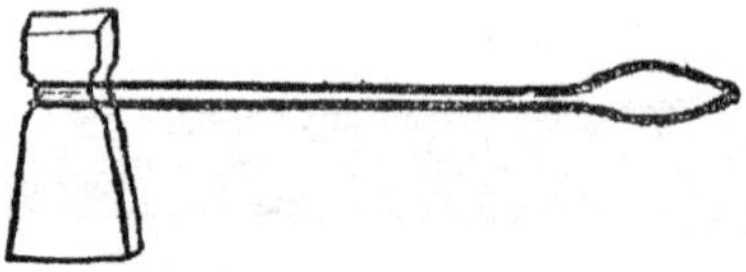

Fig. 197.—Set-hammer for Tightening Hoops.

should work with the closed socket beside him and take the measurement of the inside of the socket at both ends with a pair of inside callipers, and the size of the bulb as it progresses with outside callipers, comparing the two. The socket may have been made rather large inside, in which case it will be necessary to put on two or more layers of serving to bring the bulb to the required size. The rope should then appear as in Fig. 196. The eye of the socket must now be made red-hot.

Fig. 198.

Fig. 199.

Figs 198 and 199.—Solid Sockets with Closed Ends.

Cool down all but the extreme back of the eye A (Fig. 192) by the application of water, and open the socket out sufficiently to allow of the bulb being pushed into place. Let the end of the bulb come just to the shoulder A (Fig. 193) of the socket; then hammer the socket on the anvil, or screw it up in the vice, until it is as close as it will go. Cool it down and drive in the rivets, clinching them well, with a big strong head on

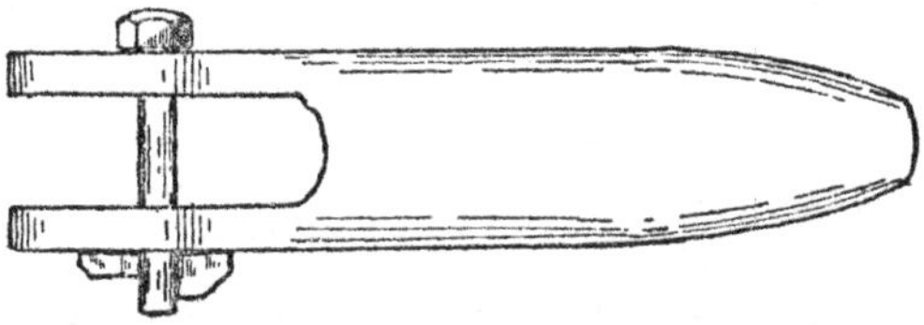

Fig. 200.—Plan of Solid Socket with Open End.

each side, for which purpose the holes must be well countersunk. The rivets should be made of good riveting iron or of mild steel. A properly filled socket should show, when finished, a $\frac{1}{8}$-in. opening between the jaws down its full length on both sides (see Figs 193 and 194), which would prove that the socket was properly gripping the bulb.

When fitting a hooped socket, first pop-mark the parts, then remove the hoops and thread them

Fig. 201.

Fig. 202.

Figs. 201 and 202.—Elevations of Two Forms of Solid Sockets with Open Ends.

on to the rope, the smallest first and the others in order; they are thus easily returned to their proper places. As soon as the hoops are replaced, the socket should be taken out of the vice and held end down on a block or anvil while the hoops are driven tightly home with a set hammer (Fig. 197) and sledge.

To give a better finish, and to prevent the hoops from working back or wet from entering,

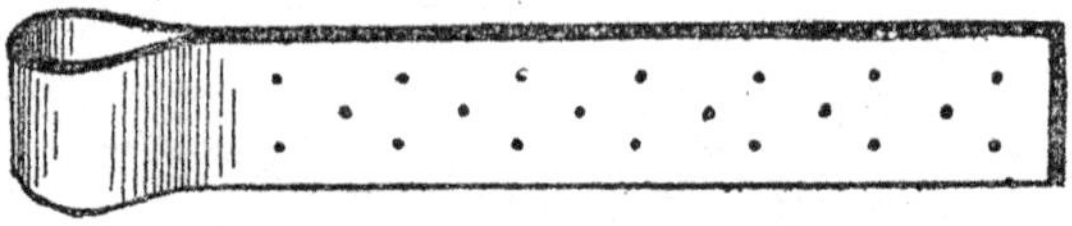

Fig. 203.—Riveted Flat Rope Socket.

they are sometimes caulked on the tapered side A (Fig. 194) with a caulking set.

For crane ropes or colliery winding ropes, steel sockets are sometimes cast solid, some with closed eyes for the D shackle, as in Figs. 198 and 199, and others with open ends with holes for a slotted pin, as in Figs. 200 to 202, the two latter being side views of alternative shapes. The form shown by Figs. 198 and 199 is the more difficult to fit, as it is obvious that the rope must first be pushed through the socket and brought out at one side of the eye before the bulb can be formed, when it has to be either pulled or hammered back until it is quite tight in the socket. After the bulb of these solid sockets has been pulled or driven into place, it is usual to drive a round tapered copper plug into the core of the bulb, to solidify it and

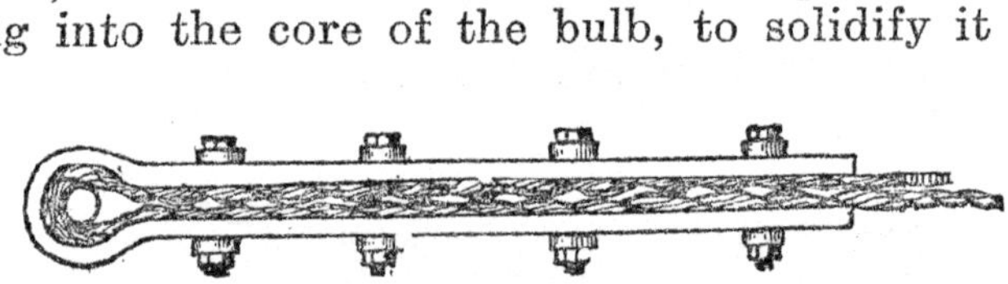

Fig. 204.—Cramped Flat Rope Socket.

increase its grip with the socket. Sometimes molten white metal is poured in to fill up the interstices between the rope and the socket. The hole at the small end of these sockets is made only a very little larger than the diameter of the rope, so that it is impossible for the bulb to pull through if properly made.

Flat wire ropes, which are not much used now, also have to be socketed, as they cannot be spliced. There are various methods of fixing flat rope sockets, varying with the strain and the

Fig. 205.—Socket Cramp.

conditions under which they work. The most usual method is to bend the wire ends back equally on both sides to make the bulb fit the inside of the socket, and to secure with a liberal supply of rivets (see Fig. 203). For stronger work, the rope-end, after being cut off square, is bent round the shackle pin, and the socket made to grip both the rope and its end, the whole being firmly secured by means of strong iron clamps and bolts (see Figs. 204 and 205). In order to drive the rivets through the compressed rope after the socket is tightened, it is necessary to make the way easy for them by driving in tapered spikes of the shape shown by Figs. 206 and 207. These require a hole in the end to take the end of a stout

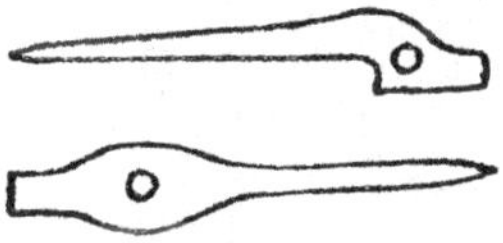

Fig. 206.

Fig. 207.

Figs. 206 and 207.—Riveting Spikes.

Fig. 208.—Rivet.

marlinespike, for the purpose of withdrawing them from the hole. They must be made of mild steel.

The best form of rivet is shown by Fig. 208; it must be made longer than the diameter of the socket, to allow for clinching the other head.

INDEX.

ALSO BY DES PAWSON

The Handbook of Knots

Pocket Guide to Knots

Des Pawson's Knot Craft

Knots: The Complete Visual Guide

Des Pawson's Knot Craft and Rope Mats

Museum of Knots and Sailor's Ropework Monograph Series

VISIT WWW.DESPAWSON.COM FOR MORE INFORMATION

Made in the USA
Monee, IL
04 November 2026

67377614R00095